D1242632

# THE 21st-century Sniper

## A COMPLETE PRACTICAL GUIDE

Brandon Webb
Glen Doherty

Skyhorse Publishing

Copyright © 2011 by Brandon Webb and Glen Doherty
All Rights Reserved. No part of this book may be reproduced in any manner without the express written consent of the publisher, except in the case of brief excerpts in critical reviews or articles. All inquiries should be addressed to Skyhorse Publishing, 307 West 36th Street, 11th Floor, New York, NY 10018.

Skyhorse Publishing books may be purchased in bulk at special discounts for sales promotion, corporate gifts, fund-raising, or educational purposes. Special editions can also be created to specifications. For details, contact the Special Sales Department, Skyhorse Publishing, 307 West 36th Street, 11th Floor, New York, NY 10018 or info@skyhorsepublishing.com.

www.skyhorsepublishing.com

10 9 8 7 6 5 4 3 2 1

Library of Congress Cataloging-in-Publication Data

Webb, Brandon.
The 21st-century sniper : a complete practical guide / Brandon Webb and Glen Doherty.
p. cm.
ISBN 978-1-61608-001-3 (pbk. : alk. paper)
1. Sniping (Military science) 2. Sniper rifles. I. Doherty, Glen. II. Title.
UD330.W33 2010
356′.162--dc22
2010034978

Printed in the United States of America

# Contents



CENTRAL ARKANSAS LIBRARY SYSTEM
LITTLE ROCK PUBLIC LIBRARY
100 ROCK STREET
LITTLE ROCK, ARKANSAS 72201

# Introduction

As a former Navy SEAL and U.S. Navy SEAL Sniper course manager, I am fortunate enough to have known and trained with some of the finest military snipers in the history of Special Operations. As the course manager, I have been responsible for the training and production of hundreds of SEAL snipers that have gone on to inflict massive precision casualties around the globe. A select few are among the most accomplished snipers in the 21st century. While I have been deployed as a sniper in the Persian Gulf, Afghanistan, and elsewhere, what is truly significant to me are the sniper students I taught that have gone on to rack up thousands of enemy kills. I consider it the greatest accomplishment of my SEAL career to have contributed to the development of the modern-day SEAL Sniper Course and its graduates. There is no greater experience for a teacher than to see his students apply their learned craft as experts and produce accomplishments greater than his own. That is the best compliment a student can give a teacher.

There are many books on the subject of snipers, but very few were written by the actual experts themselves. As a voracious reader, I

**This is a great example of a 21st-century shooter/spotter pair on display. This is a long-distance and applied-technology configuration.**

frequent many bookstores, and it was frustrating to see the massive amount of books written on the topic of sniping by people who have no experience on the subject. This led me to write *The 21st-century Sniper*. After all, would you take medical advice from someone who has never practiced the art of medicine? I am very fortunate to have a great friend and former SEAL sniper classmate, Glen Doherty, as co-author and contributor on this project. This book would not have been finished if it were not for Glen's diligence and the occasional sharp kick in the ass he gave me. Glen and I were not only classmates, but we were paired together as a shooter/spotter team throughout our initial SEAL sniper training. Glen is a gifted shooter and a natural on the trigger and down the sights.

This book is written by actual snipers who have seen combat in Iraq, Afghanistan, and elsewhere. We hope to shed light on the unique and often misunderstood community of these highly intelligent, patient, and skilled marksmen. The sniper's role on the battlefield is a very personal one, but it is also very effective. Snipers are highly skilled operators on call to instill terror into the hearts and minds of the enemy. They are employed in a very specific way to ensure no collateral damage associated with the kill. This fact alone makes snipers one of the most effective tools in a military commander's arsenal.

Our goal with this book is to represent the sniper community in a positive way and take the reader on a rare personal journey behind the scenes from an actual sniper's perspective. We take you back in time to reflect on the world's first snipers. We discuss in detail the attributes and life experiences that make a great sniper. Then we journey forward and provide a rare look at some of the best weapons and tools of the trade in use today. Technology and advanced training methodology are a game changer in the new century.

While taking special care not to give away secrets of the trade, we discuss the latest in training methodology, weapons, optics, lasers, and kit. Then we provide a look at what's to come in the next decade. We also give actual stories from our personal archives as well as stories from some of the most accomplished snipers of the 21st century. While the names and specific locations remain nondescript, these are in fact true modern-day stories from actual snipers employed in the current complex environment of asymmetrical warfare.

The lone sniper's ability to inflict death and terror with precision similar to the uncaught serial killer can cause immense devastation on the battlefield, defeating both the body and mind of the enemy.

A book like this has never been written before. Enjoy!

1

# The History of Snipers

"Certainly there is no hunting like the hunting of man, and those who have hunted armed men long enough and liked it, never really care for anything else thereafter. . . ."

**—Earnest Hemingway from "On the Blue Water," *Esquire*, April 1936**

How can you truly prepare for the future without knowing the past?

Bring up the topic of snipers at your next dinner party and see where the conversation goes. It is a topic full of controversy and mystery, conjuring a range of attitudes and images depending on the individual. Hollywood has done its share to contribute to this fascination: An entire generation of modern youths will grow up having played realistic games on computers simulating combat sniper operations. Ask a middle-aged mother from New England and you might get a recap of an old *Cheers* episode where Sam Malone and his friends lure Dr. Crane out to the woods for an old-fashioned "snipe hunt," an old-school practical joke, which leaves the mark deep in the woods with a sack, making clucking noises and hoping the elusive snipe will come jumping into the bag. The darker connections to the term will remind some of the Beltway Snipers: John Allen Muhammad and Lee Boyd Malvo terrorized the D.C. area in 2002, shooting innocent victims and later getting caught after one of the largest manhunts in modern history. Veterans of the armed forces might remember quiet men who kept mostly to themselves and carried scoped weapons and disappeared into the night, sometimes not returning for days on end. Whatever the term "sniper" brings to mind, it has a long history. In the following chapter we'll trace its roots back to the early days of marksmen, riflemen, sharpshooters, and hunters that knew the power and potential that one well-aimed shot could have.

Without going all the way back to prehistoric times, there are some significant historical events that have changed the modern world and weapons as we know them. The projectile played an important role in warfare since the beginning of time. Slings, spears, bows and arrows, crossbows, then later muskets and rifles, are all tools with specific applications, relevant for their particular time periods. The bow begat the crossbow, which dominated the battlefield during the Middle Ages and was so effective that Pope Innocent II decreed the weapon "unfit for combat amongst Christians." This edict didn't apply to combat with Muslims, however, so when King Richard the Lionheart took his army toward Jerusalem during the Third Crusade, he was well equipped with crossbowmen. His smaller force held off many attacks from Saladin's larger forces, thanks to the ability of King Richard's bowmen to maintain an accurate and rapid rate of fire. Ironically, it was when Richard was returning to his homeland from his failed crusade that he was struck down by the very weapon his Pope had sought to ban. In 1199, King Richard was hit by a crossbow bolt from the ramparts of a castle he had under siege in Limousin, France, and later died of the wound. His death

Early 17th-century Wheel-lock Pistol from England

was the result of a medieval sniper conducting what would later become common practice on the battlefield: directly targeting leadership to affect command and control and demoralize those still alive.

The Chinese had been using gunpowder since the ninth century, but it didn't really become important to European superpowers of the day until much later, entering Europe most likely through Arab trading routes in the thirteenth century. Initially it was employed mostly in large cannons and siege engines, monstrous machines that hurled massive balls designed for knocking down walls. During the next several hundred years technological improvements were made to both the powder and the weapons it was used in, developing from knowledge gained by the many who fought around the world at the time. Ultimately, the flintlock musket would become a mainstay of the British and European Armies, followed by the matchlock, and for some, an invention inspired by Leonardo da Vinci himself—the wheel lock firing mechanism.

The difference between a rifle and a musket is that a musket has a smooth bore, whereas in a rifle, "rifling," or spiraling, grooves are cut into the inside of the barrel, creating a spinning motion on the departing projectile. This motion stabilizes the projectile in the air, allowing it to fly farther with a much greater degree of accuracy. The gyroscopic stability of a spinning projectile had been known since the time of the bow and arrow and crossbow, and fletching was specifically placed on arrows and bolts on a cant to impart spin and improve accuracy and range. It didn't take long for gunsmiths to apply this same knowledge to modern weapons, but in the early stages, the musket and rifle had separate and specific applications.

The reason the smoothbore musket remained the primary weapon for infantry at the time was its ability to elicit a higher rate of fire from a trained soldier. In a musket, the ammunition was made to be slightly smaller than the diameter of the barrel, so in reality the accurate range was most likely less than one hundred yards, as the ball would come rattling out of the barrel. For a ball to engage the rifled grooves of the barrel, it had to be large enough

Rifling exposed (from a 105mm tank gun)

to span the entire diameter and, because of this, took longer to load. In addition, the rifled grooves were prone to fouling from the gunpowder, so often that the barrel would have to be cleaned between shots. Fortunately for military tacticians of the time, volley fire was the flavor of the day, engaging large quantities of massed troops in a line shooting into other massed troops, one line firing, the other loading, not having to aim at a specific individual but at a mass of soldiers who would still be visible at close range even through all the smoke. After several rounds of volley fire, a bayonet charge could be expected. Put yourself there for a moment: You're shoulder to shoulder, wearing a wool jacket and hat, sweating like a pig under the hot sun. Your heavy musket is loaded and your hands are sweaty from nerves and the heat. Your mouth is dry; cannons fire from your front and rear; and as you march forward lines to your left and right are decimated by artillery fire. Smoke hovers over the field, fires burn in the distance, and men can be heard screaming in pain in all directions. Your ears are ringing, yet you maintain your cadence, as you have been trained to do, taking direction from the officer to your flank. Shortly, a blur of color appears through the haze, and a halt is called. The command to make ready is heard, and you cock your weapon and stand at the high port. You hear a similar call being made across the field. "Take aim." You point your weapon toward the mass of troops. "Fire!" Loud claps of thunder seem to engulf the entire area, flashes from the gunpowder igniting in the pans and barrels erupt, and smoke fills the air. Men drop to your right and left. Blood is everywhere, covering you as you seek to remember the natural rhythm of a reload. Measure the powder down the barrel, a different amount to the flashpan, place your ball and wadding in, ram it home, recock, and ready again.

To a modern-day soldier, especially a well-trained sniper, these tactics of old seem insane. How could a sane person stand at such close range and receive round after round of fire from an enemy, without at the very least lowering his profile by getting in the prone position or seeking some type of cover? The discipline of the troops and their dedication to duty, their fellow soldiers, and their pride and love of country truly must have been incredible. It is hard, if not impossible, to imagine. There is a modern-day military expression that comes to mind: "If you're going to be stupid, you'd better be tough." Sounds about right.

Rifles were common at the time, but were usually more expensive, and most likely used in the country for hunting and by marksmen for sport and competition. With the discovery of the New World and the rapid colonization of new land by the European superpowers, the rifle would prove invaluable for settling new and inhospitable country and would force the Old World military tacticians to reevaluate their methods. The Germans had developed the leading rifle of the day, commonly called the Jaeger (hunter) rifle. Its short barrel fired a .50 caliber or larger round and was quite accurate for its day. Many of these weapons found their way to the colonies with German settlers, where in the heavily German area of Pennsylvania the gun evolved into what is now called the Pennsylvania Long rifle, or the Kentucky rifle. (They were later named this because they were used by the Kentucky sharpshooters against the British at the Battle of New Orleans during the War of 1812.) Life in the New World was hard: there was more land and more open space, meaning longer,

**Volley fire**

more accurate shots had to be made in order to put meat on the table. The hunters and woodsmen of the day had to adapt to their environment, blend in, learn the art of camouflage, and develop stalking skills, patience, and the ability to have one-shot success at longer ranges in order to keep their families fed. All of these skills would come into play in the wars to come and are still used today by the modern military sniper.

Major Robert Rogers

Prior to the American Revolution, several conflicts in the New World gave rise to evolving tactics, weapons, and a new type of soldier—the ranger. These men were woodsmen and hunters who operated in small units outside the safe confines of villages and forts. They engaged hostile Native Americans and European forces that encroached on the new colonies. The skills inherent to the success of these units were later passed on to the next generation, prior to the onset of the American Revolution. The men traveled light and fast, shedding weight and excess gear wherever possible, and adopted fighting and raiding tactics of the Native Americans they fought against—guerilla warfare in its earliest forms. One of the early students to receive the skills of these early Native

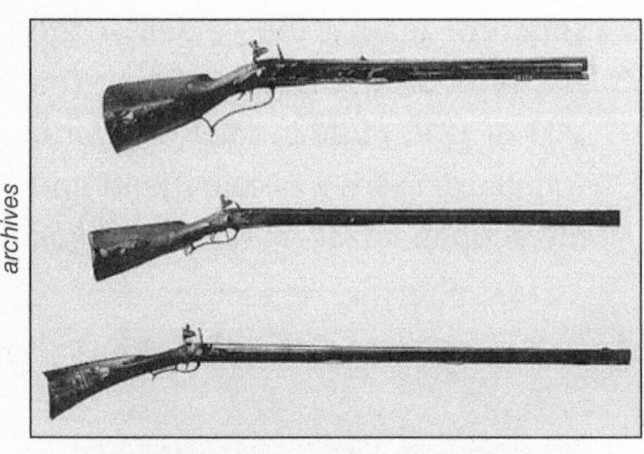

Photo Credit: University of Virginia archives

Pictured at the top is a Jaeger rifle made by Andreas Staarman in late-seventeenth-century Berlin. It has a 26-inch barrel and .75-caliber bore. Next is a New World hybrid owned by Pennsylvanian Edward Marshall in 1737. Like the Jaeger, it has a patch box with a sliding wooden cover. At the bottom is an early example of a fully evolved American long rifle. Its 44.5-inch barrel made it more accurate, and its .44-caliber bore made it more efficient in its use of lead and powder, than the heavy Jaeger.

Photo Credit: 2010 Fort at No.4 Educational Archive

**"The Advanced Guard" June 21, 1759. Based on primary source descriptions of the campaign, this picture depicts Major Robert Rogers and "Captain Jacobs" as they scout the forest ahead of General Jeffrey Amherst's army on their way to capture Fort Carillon (Ticonderoga) and other French posts on Lake Champlain.**

American hunters was a man named Robert Rogers, later to become Major Rogers who led the famous Rogers' Rangers, formed in 1756 during the French–Indian wars.

*Rogers' Rules of Rangering* is an illuminating read that shows the difference in tactics relative to the conventional troops of the day:

> *Rogers' Rules of Rangering, est. 1754*
>
> Don't forget nothing. Have your muskets clean as a whistle. Tomahawks scoured. Sixty rounds powder and ball and be ready to march at a moment's notice. When you are on the march, act the way you would if you were sneaking up on a deer. See the enemy first. Tell the truth about what you see and what you do. The entire army is depending on you for information. Don't lie to a ranger or an officer. Never take a chance you don't have to when we're on the march we march single file far enough apart so one shot can't go through two men. If we strike swamps or soft ground, we spread out abreast so it is hard to

track us. We move till dark; when we camp half the party sleeps while half the party stays awake. Don't ever march home the same way. Take a different route back so you won't be ambushed. Every night you will be told where to meet if surrounded by a superior force. Don't sit to eat without posting sentries. Don't sleep past dawn. The French and Indians attack at dawn. If you're being followed circle round your own tracks and ambush them. When the enemy is aiming at you kneel down or lie down. Let the enemy come close enough to touch then let him have it. Then jump up and finish him off with your tomahawks.

Many of the principles above are still very much in practice by SEAL snipers today. If it ain't broke, don't fix it.

## American Revolution

During the American Revolution, British regulars were armed with what was called the Brown Bess, a smoothbore flintlock musket, while the German Hessian mercenaries carried short-barreled Jaeger rifles. The colonists used an assortment of weapons, including the Brown Bess and the very accurate Pennsylvania, or Kentucky, rifle. It was the first conflict where sharpshooters were widely used, and the beginning of deliberate targeting of officers and other high-value targets on a regular basis. Two parallel stories embody the outcome of the war and the role of snipers during that war—the famous "shot that was never taken" and the shot that was.

In the first, an inventor and marksman from Britain named Patrick Ferguson, then a Captain of a unit of sharpshooters, was armed with his namesake: the breech loading Ferguson rifle. This rifle was able to maintain the same or greater rate of fire as a smoothbore musket, but without the common problems with fouling and reloading (the Ferguson rifle could be reloaded

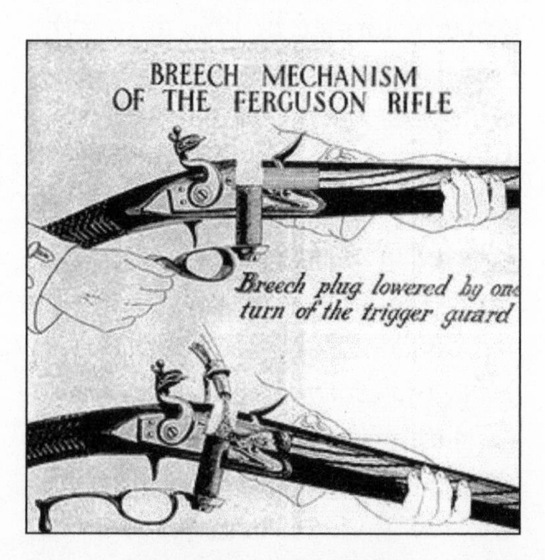


Open breach Ferguson rifle

while on the move) that had plagued earlier attempts, and with amazing accuracy. Prior to the Battle of Brandywine in September of 1777, while in a hide-site along a river, Ferguson and several of his soldiers witnessed an American officer and his French Hussar cavalry companion ride within 100 yards, stop, and take into account the surrounding area. Initially Ferguson ordered his men to dispatch of the two-easy shots with the Ferguson rifle—but upon second thought he rescinded the order, considering it un-gentlemanly to deliberately target fellow officers who were not directly bearing arms against them. In the pitched battle that followed, Ferguson was injured, and while recovering learned that the target he so chivalrously allowed to pass was General George Washington himself.

In direct counterpoint to Ferguson, the Colonials had Daniel Morgan, a talented tactician and marksman—rumored to be a distant cousin of Daniel Boone, the famous frontiersman and Indian fighter—who led specialized groups of shooters. Morgan's Sharpshooters (We prefer Morgan's Marksmen) initially made a 600-mile forced march from Pennsylvania to Boston to engage the British during the initial stages of the war.

Photo Credit: *Legacy Journal*

**One of Morgan's Sharpshooters armed with a Pennsylvania Long Rifle, later known as the Kentucky Rifle.**

Armed with Pennsylvania long rifles, his 11th Virginia Regiment of 400 men were engaged in a fierce battle in October of 1777 at the Second Battle of Saratoga. General Fraser led one flanking unit of the British forces. His troops were under heavy fire from Morgan's rifle corps, suffering devastating casualties and the disintegration of order and discipline. Leading from the front, Fraser galloped along his lines bolstering the morale of his troops and directing a counteroffensive. It was none other than General Benedict

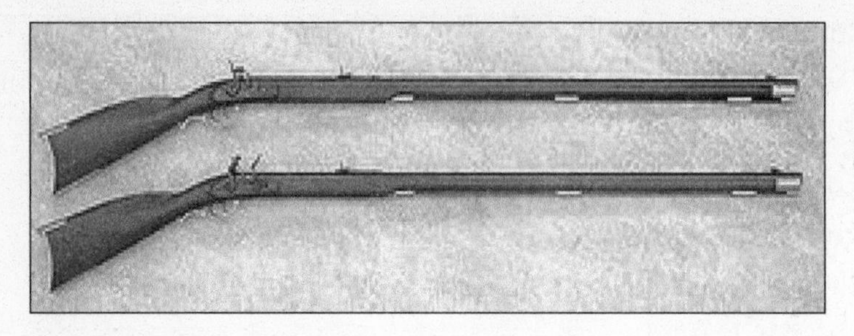

The Pennsylvania Rifle became known as the Kentucky Rifle after the Battle of New Orleans during the War of 1812 when a regiment from Kentucky used them to great effect against the British.

Arnold who gave the command to Colonel Morgan to target Fraser and bring him down immediately. Several shooters were most likely tracking and firing at Fraser shortly after the order was given, but a young sharp-shooter named Timothy Murphy, perched in a tree and shooting at a range of approximately 300 yards, was the one credited with the kill. With their leadership gone, the British and Hessian troops folded, and the Colonials won the day. It was a huge victory for the Revolution, one that was celebrated even across the pond in Paris, France, an enemy of Britain at the time. Both Ferguson and Morgan would find their way back into pivotal battles during the Revolutionary War. Ferguson met his end at the Battle of Kings Mountain—another huge victory for the Continental Army, and Morgan, later promoted to general, won a resounding victory at the Battle of Cowpens against Colonel Tarleton, where he allegedly told his men, "Aim for the epaulettes." Morgan continued to promote the use of sharpshooters throughout the remainder of his career and was instrumental in modifying future long rifles to be fitted for bayonets, one of their major weaknesses exploited by the British during the War of Independence.

In 1929 FDR dedicated a statue in New York State to Timothy Murphy, and said the following during the ceremony, which should resonate with soldiers everywhere:

> This country has been made by Timothy Murphys, the men in the ranks. Conditions here called for the qualities of the heart and head that Tim Murphy had in abundance. Our histories should tell us more of the men in the ranks, for it was to them, more than to the generals, that we were indebted for our military victories.

With the turn of the century there was the rise of a vertically challenged

**60th and 95th Riflemen shooting the Baker rifle**

little man who brought with him a tactic that had yet to be used on a grand scale until his push across the European continent. Napoleon Bonaparte firmly believed in what would later be called blitzkrieg warfare: fast-moving, independent units rendering death and destruction, maneuvering quickly, operating independently, and wielding accurate rifle fire to its utmost potential. To counter Napoleon, the British sent in the newly formed 95th Rifles and the 60th Regiment of Foot (later named the Rifle Brigade and then the Kings Royal Rifle Corps), a brigade of the finest shots the crown could muster.

Using the new Baker rifle, these units went head-to-head with Napoleon. It's been said that the modern sniper expression "one shot, one kill" can be attributed to these units of marksmen. These units had foresworn the chivalrous code of battle from the old days and now specifically targeted officers, bugleman, drummers, and artillerymen. Many battles fought by these new units produced casualty ratios as high as one officer for every five infantrymen killed, an astounding number considering the structure and tactics of standard military units. Once again the effectiveness of quality shooters proved their worth in the field.

A small victory for the French, despite Napoleon's eventual defeat, came at the hands of a sniper posted in the rigging of the French ship *Redoubtable*. In October of 1805 Admiral Lord Horatio Nelson was in command of HMS *Victory* and

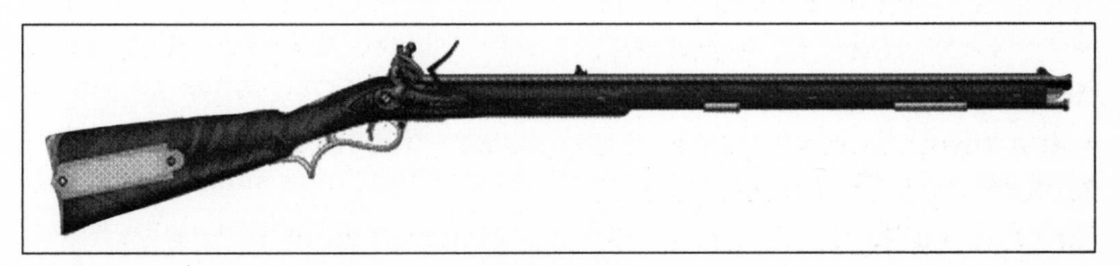

**Baker rifle**

**The death of Admiral Nelson**

engaged in a battle at close quarters with the French. It was a common tactic for the French, Spanish, and Portuguese during that era to send snipers aloft into the rigging to kill officers and take out gun crews. The British, allegedly due to Admiral Nelson's strategies, hadn't opted for those tactics and really hadn't needed to, as Nelson had successfully crushed any naval opposition up until that point. From a range inside 100 yards, Nelson was brought down by a French sniper, and despite the British winning the day, they lost their most seasoned naval commander.

Leading up to the Civil War, there were several technological advances that would change the way battles *could* be fought, as well as vast improvements in the reliability and range of the modern rifle. One of the first significant innovations was the development and implementation of the percussion cap, as opposed to the old flintlock system. We've all heard the expressions "flash in the pan" and "hang fire." These both come from the antiquated and unreliable flintlock firing systems, where the flint ignited powder in the flashpan, sending a flame through the touchhole into the barrel. A "flash in the pan" would occur when the touchhole was clogged, and only the primer charge would ignite. Flash, but no bang. A "hang fire" would occur when the main charge wouldn't ignite initially, but there might be a small spark still smoldering in the touchhole, which in a second or a minute might send a bullet out the barrel. Not

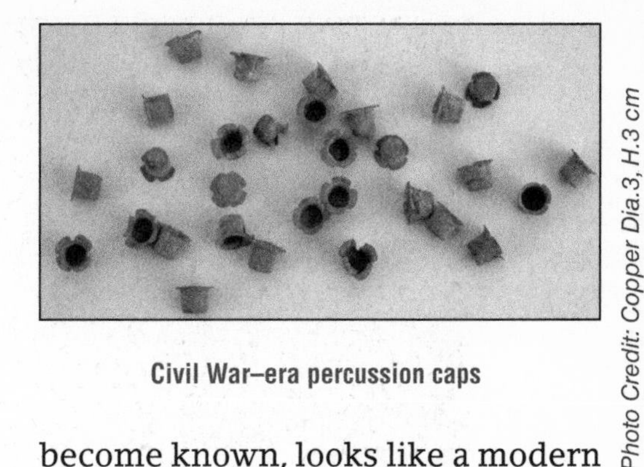

Civil War–era percussion caps

Photo Credit: Copper Dia.3, H.3 cm Vicksburg National Military Park, VICK 34, Carol M.Highsmith, photographer

good. Another thing the percussion cap did away with was the long "follow through" that was required, meaning keeping your sight picture and alignment perfect for several seconds as the successive charges finally sent the ball out of the barrel. This was easier said than done when holding a long, heavy weapon and taking effective fire from the enemy. With the percussion cap, misfires became extremely uncommon, and wet weather became almost irrelevant to loading and firing. The small cap was about the size of an eraser head and was placed on top of a nipple with a pinhole that led directly into the barrel. When the hammer hit the cap, the flame was directed through the hole of the nipple straight into the barrel, which would ignite the main charge. Many of the flintlock rifles of the day were converted so that they could be used with the new percussion cap ignition system.

The next major breakthrough would change ballistics forever. In 1847, Captain Claude Minié redesigned a round that had originated but was not adopted in Britain. The Minié ball, as it would become known, looks like a modern bullet, a conical-cylindrical shaped self-expanding lead bullet. The original design had an expanding iron cup in its base that would be driven up into the lead and force it to engage the rifling grooves.

Too often the iron cup would punch right through the lead, sending a mess of shrapnel out the end of the barrel. Fortunately, an American named Burton perfected the design

The percussion hammer ready to strike.
The cap would be placed on the nipple.

Photo Credit: Brain, Marshall "How Flintlock Guns Work." 01 April 2000. Howstuffworks.com

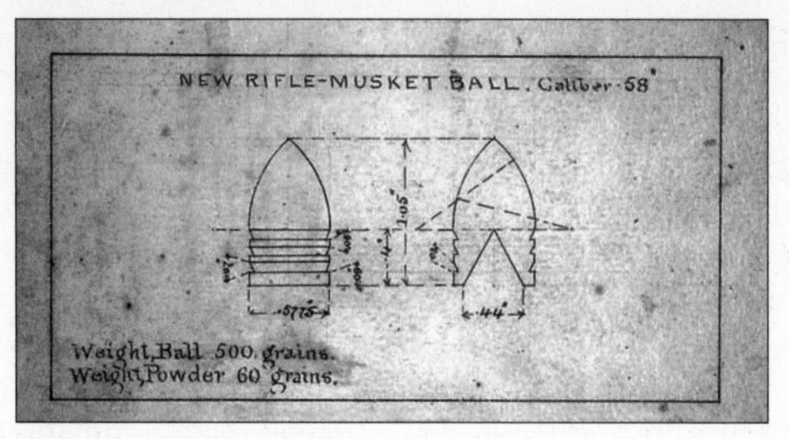


Diagram of mini ball design

(do we have to do everything?) but never got the name recognition, and the round forever to be known as the Minié ball would be taken into the bloodiest conflict the United States has ever seen. The results of this new round were immediately nothing short of phenomenal. Now the average soldier was capable of accurate fire out past 300 yards and volley fire past 1,000. Unfortunately, the brilliant tacticians leading troops during the Crimean War and later during the Civil War still enjoyed massing troops shoulder to shoulder and exchanging fire at close range. I try to imagine a modern-day SEAL platoon being ordered to form a line and shoot offhand (standing) at another platoon a couple hundred yards away. I can imagine that whoever delivered that order might be the first one shot and not necessarily from the other side!

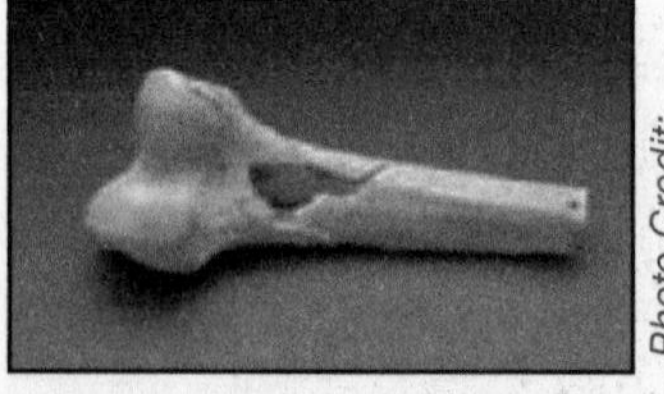

*Photo Credit: National Museum of Health and Medicine, Armed Forces Institute of Pathology, Washington, D.C.*

Human femur shot with a 510-grain lead Minie ball fired from a .58-caliber Springfield Model 1862 Rifle.

## The American Civil War

Both the Union and Confederacy effectively used sharpshooters during the war, and several generals from both sides suffered the same fate as one General Sedgwick, whose last words before being shot in the face by a Confederate sniper from a range at 800 yards were, "They wouldn't hit an elephant at this distance." The Industrial Revolution was in progress, and the Springfield Armory in Massachusetts produced over one million rifles

Photo Credit: Library of Congress

The death of General Sedgwick

for the Union. This reliable and effective rifle was truly a modern weapon for its day, with front and rear flip-up leaf sights and maximum effective range out to 500 yards. The one real drawback to the Springfield was that it was still loaded from the muzzle, and the only easy way to muzzle load quickly was standing up—not something one wants to do with Confederate marksmen in the field.

The Union had someone else pushing the envelope for battlefield deployment of trained shooters: a man named Hiram Berdan. Hiram was an engineer, an inventor, and a crack shot. He also was a wealthy man and had the ear of the president and secretary of war. He convinced them to allow him to recruit what would later be known as Berdan's Sharpshooters, two regiments formed by an active recruiting campaign throughout the North, with each candidate having to pass a rigorous marksmanship test.

The rifle that would eventually land in the hands of Berdan's troops was the famous Sharps rifle—a

General John Sedgwick's Corps, waiting for orders, 1863

**Berdan's Sharpshooters working in an early shooter spotter pair with security element.**

**Harper's Weekly, October 5, 1861.**

Dead Confederate "sharp shooter" at Devil's Den. This photo was later found to be staged by the photographer.

edge would most likely have gone to them, because more of their troops came from rural communities, where being an accurate shot made the difference between eating and going hungry. Unfortunately for the South, they lacked the industrial complex that the North had, and despite happily picking through the dead and acquiring Springfields and Sharps, the rest of their rifles were either brought from the farm or imported.

Despite significant success by both sides during the Civil War, the powers that be still didn't embrace a philosophical change in tactics that would have more thoroughly used the

breech-loading rifle capable of incredible accuracy and reliability. It was often specially configured for Berdan's soldiers with a double trigger, the first to set, the second a hair trigger. Berdan himself was a bit of a boob, not respected as a combat commander, and spent most of his time schmoozing in D.C. as opposed to being in the field. His sharpshooters distinguished themselves throughout the war and produced plenty of impressive long-range kills of senior officers, artillerymen, NCOs, and regular troops.

The Confederacy wasn't without its own skilled shooters. At the beginning of the war the

Dead Confederate soldier in the trenches at Ft. Mahone, Petersberg, VA.

*Photo from: http://www.taylorsfirearms.com/products/bpSharps.tpl*

**The British Enfield P53 rifle was imported to the South early on, as well as the incredibly expensive and accurate Whitworth rifle. This process became harder and harder as the North's Naval blockade became more efficient and effective. The Whitworth was a muzzle loader, with a hexagonal barrel that required special ammunition, and is said to have been purchased the base model for $600, and the top end which include a telescopic scope and 1000 rounds of ammunition for $1000. That's a lot of money in the 1860's, close to $20,000 in 2009 currency. This weapon could reach out to 1800 yards and was only given to the best of the best shooters from the South, and is credited with multiple high value kills, including that of the aforementioned General Sedgwick.**

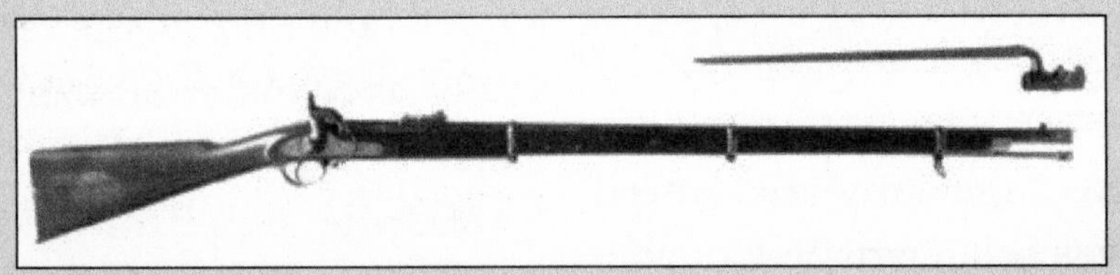

*Photo and caption from: http://www.cfspress.com/sharpshooters/arms.html*

**Enfield P53 rifle. "Confederate ordnance chief Josiah Gorgas called the Enfield, or British Pattern 53 Long Rifle-Musket, named for the year of its adoption, "the finest arm in the world." Sturdy, reliable, and extremely accurate even at extended ranges, it consistently outshot everything but the Whitworth and quickly became a favorite on both sides. This nine-and-a-half-pound, single-shot, muzzle-loading, .577-caliber rifle was as close to a standard infantry weapon as the Confederacy ever got and was also used in large numbers by the Union."**

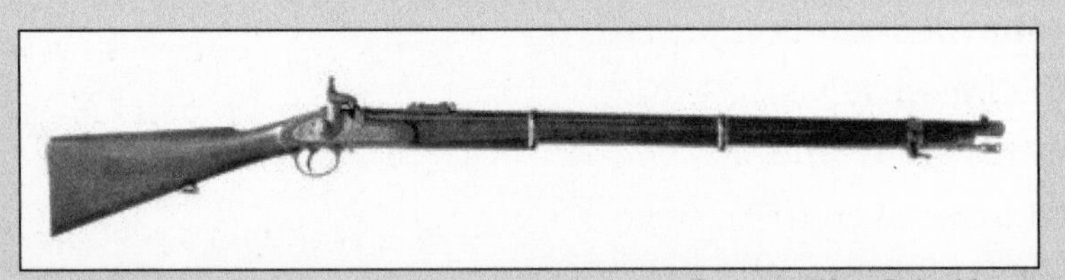

*Courtesy West Point Museum*

**The Whitworth rifle fired a unique hexagonal bullet that sported a muzzle velocity of approximately 1300 feet per second. It was not as good of an all around infantry weapon as the Enfield, but as a long range shooter its accuracy was unparalleled for the era. It could be fitted with a bayonet, and a variety of sights, including a four power telescopic sight. "The claim of 'fatal results at 1,500 yards,'" concluded one modern expert, "was no foolish boast." Overall, it was a deadly weapon that, in the right hands, repaid its high cost many times over. "I do not believe a harder-shooting, harder-kicking, longer-range gun was ever made than the Whitworth rifle," asserted sharpshooter veteran Isaac Shannon. (courtesy West Point Museum)**

special skills of these early breeds of snipers. Berdan's units were issued green and brown uniforms, and the Confederates were known to employ some early styles of ghillie suits, pinning leaves and brush to their clothes while stalking Federals. Still, this bloody war pushed technology to the limit, and a lot of the equipment and tactics of the modern sniper can be traced directly to the War Between the States.

## Leading up to the first Great War

Excellent, if somewhat little-known, examples of the importance of sharp-shooting in battle are the First and Second Boer Wars. The Boers were farmers, descendants of Dutch and German settlers living in two distinct areas, the Transvaal and the Orange Free State. The date was December 1880 and for the next eight months the Boers thoroughly embarrassed the British troops. The Boers fought in a hit-and-run style with breech-loading rifles against the scarlet-clad British troops who, despite extensive combat experience, still hadn't adopted tactics that were able to counter the Boers' accurate shooting and rapid movement. The Boers, similar to the Confederates, were born with rifles in their hands

During the first Boer War, the British took a pounding.

and were experienced scouts and horseman with a lifetime of field craft and local knowledge. During multiple battles, the Boer marksmen would use a relatively new technique of "fire and movement," where one group would keep the Brits' heads down while the other would maneuver to a more tactically advantageous position, then switch. It didn't take long before the British withdrew, and the Boers of the Transvaal declared independence once again. But the British weren't done yet, and after licking their wounds, they would return with new weapons, new uniforms, new tactics, and a new resolve.

## The Second Boer War

This war began in October of 1899. For a decade prior, British citizens had been flooding into the area to seek their fortunes in gold, which was recently discovered there. The peace between the two groups had never been solid, and the Boers were ready when the second conflict erupted. They had been purchasing large quantities of the bolt-action German Mauser rifle, which used recently perfected smokeless ammunition and was capable of holding ten rounds in an internal magazine. The British had adopted the Lee-Enfield, which was also a bolt-action rifle with an internal magazine but still used traditional black powder charges. The development of smokeless powder was a tremendous technological coup for gun and ammunition manufacturers, and heralded the beginning of the true sniper era. Prior to this, even if firing from cover and concealment, a sharpshooter was going to give away his location by the large plume of smoke that would erupt from his weapon after firing. Now there was only the muzzle flash, which can be very difficult to detect during daylight hours. The Boers "went Commando," organizing into small mobile units with democratically elected commanders and NCOs. They were totally self-sufficient, procuring their own food, water, and shelter as they harassed and killed the British. In one particularly bloody battle at a hill called Spion Kop (also referred to as the Acre of Murder), the British suffered some 1,500 casualties. In one shallow trench on top of the hill, 75 British soldiers lay dead, each with a bullet hole in his head,

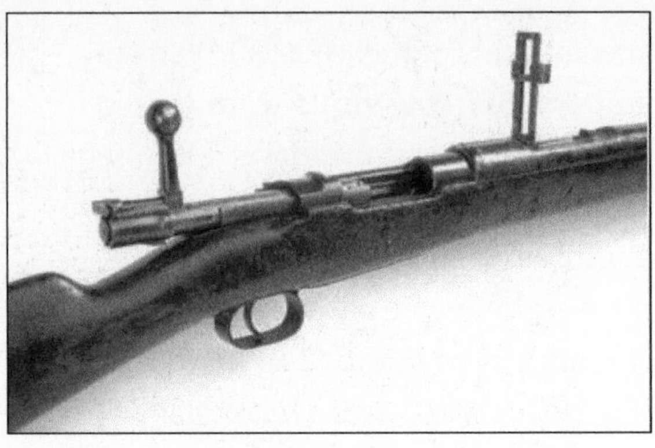

Mauser rifle

Boers engage in some long-range shooting.

Photo Credit: Hillegas, Howard. With the Boer Forces, eBook 16462

**Battle-hardened Boers pose for a rare photograph.**

victims of the rapid-firing Mauser and the Boer snipers.

Ultimately it came down to sheer numbers, and despite the ability of the Boers to evade capture and continue their guerilla campaign, a peace was negotiated in 1902, with shared control of South Africa going to the Boers and British.

## World War I

When the Germans failed to quickly overrun France in 1914, what was then called the Great War quickly settled into a prehistoric war of attrition. Trench warfare was not a new concept, but both sides used new technologies or innovations such as mustard gas, hand grenades, new artillery shells, machine guns, and, of course, snipers to great effect in the slaughter. The term "sniper" is said to come from India, where British officers hunted a small bird called the snipe, which is much like the North American woodcock. The snipe was quick, hard to see, and flew in erratic patterns. You had to be quite the marksman to shoot this bird, and those who developed a certain level of prowess in the sport were called snipers. Wherever the name came from, it came into common use during World War I. For the first couple years of the war, the German snipers dominated the Western Front, and British and French newspapers would commonly relate the horrific stories of life in the trenches and how quickly the accurate fire of the German shooters could end a life. Letters from British soldiers on the front were filled with stories of German snipers, telling tales of fellow soldiers taking the briefest of looks over the berm and falling backward, blood spilling out the back of their head. The Germans were quick to appreciate the powerful effects snipers could have on lines that often were less than 400 meters apart and equipped their best marksmen with some of the most technologically advanced rifles and optics in the world. Life in the trenches was hard enough—especially since the Germans had secured most of the high ground, with better drainage in their trenches, whereas the Allies' trenches were in the low ground, full of water and mud, vermin and death. Initially, the Germans also cut their trenches in a sawtooth pattern and used plenty of armor plating with "loopholes" to shoot through, whereas the Allies' trenches were mostly straight, forcing the soldiers to actually step up and set up on top of the meticulously level berm. The French and British

**Gun crew from Regimental Headquarters Company, 23rd Infantry, firing 37mm gun during an advance against German entrenched positions. U.S. 2nd Div., St. Mihael region.**

**A life in the trenches during WWI**

eventually learned, and things did change, but not quickly.

It took the British and French a bit too long to realize the need to counter the German snipers, and there are several men worth mentioning that were the catalysts for the Allied forces to finally train and produce their own snipers. First and foremost, Major Hesketh-Pritchard, who initially tried to get into the war at the age of 38, but was denied, yet persevered by getting to the front as a war correspondent. He had an extensive background in big-game hunting and eventually pushed the need for Allied snipers up the resistant and overly bureaucratic chain of command. He was later commissioned in the British Army and returned to the trenches where he and his unit went up and down the line, training soldiers in how to use their scoped rifles that had been indiscriminately handed out. What he found wasn't unexpected: no one was trained properly and most of the scopes were out of alignment or hadn't been zeroed. It was a start, anyway, but he wasn't through yet. Eventually he got the high command to allow him to set up the very first sniper school, called the First Army SOS School (sniping, observation, and scouting). It became a 17-day course, covering much of the same curriculum as any modern-day sniper course would teach: rifle and scope setup and maintenance, ballistics, tactics, camouflage, obser-

Photo Credit: National library of Scotland

**Flooded trenches along the Western Front. Looks like they are having a good time.**

vation skills, countersniper techniques, and, of course, marksmanship.

He also established the shooter/spotter pair and believed, as I do, that it is actually much more difficult to be the spotter than the shooter. The initial course was in Bethune, France, but was later moved to England, and was so successful that shortly thereafter a second army school of sniping was created.

Another British Army unit that contributed extensively to the newfound snipers' role in World War I was the Lovat Scouts, mostly formed from Scottish Highlanders. Many of these men had been gamekeepers in their previous lives, often called ghillies (from the Gaelic for "servant," as they would guide hunts for the rich) and had spent their days tracking animals, reading maps, observing, stalking deer, and practicing the art of camouflage. The modern-day ghillie suit traces its origins back to these men. They would don loose fitting robes and use local vegetation, which was woven into their suits to help them blend into their surroundings and effectively disappear. The ghillies perfected the art of camouflage, and their lessons were later taught at the first and second sniper schools, which would be attended by Americans after they joined the War in 1917.

Photo Credit: Sniping in France by Major H. Hesketh-Prichard

**A contrast showing the drawbacks of a uniformed shooter versus the advantage of a good camouflaged shooter.**

Photo Credit: www.nps.gov

**A soldier in WWI uses a Springfield armory modified rifle aimed through a periscope device.**

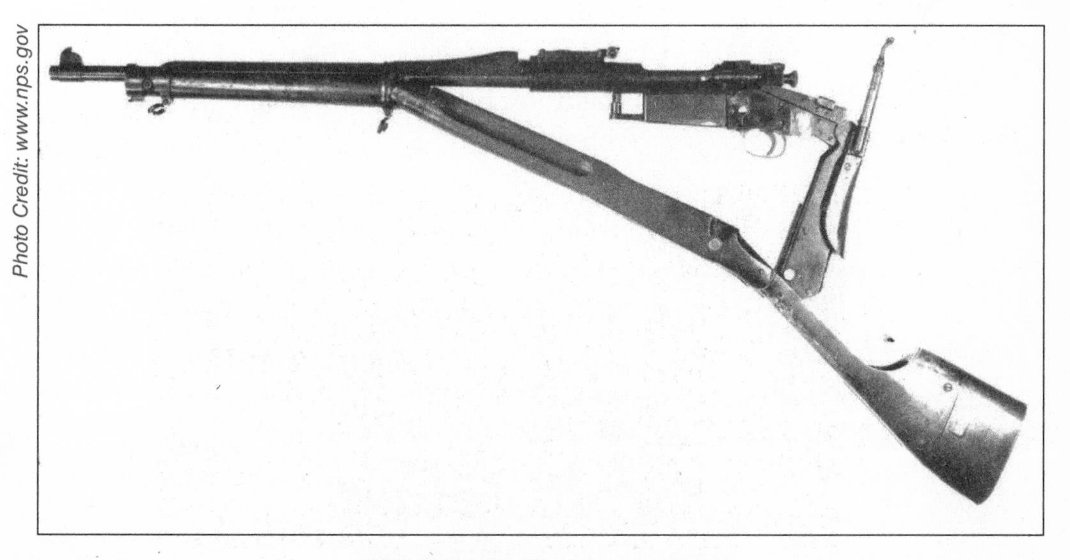

Photo Credit: www.nps.gov

**1903 Springfield with periscope attachment**

"Brandon and I were shooting partners in sniper school and managed to post the highest score for Navy personnel during the seven-week shooting phase of the course. We always managed higher scores with me on the trigger and him as the spotter. The guy was gifted on the spotting scope, able to accurately call wind and adjust for environmentals no matter the conditions. All I had to do was break a clean shot and have a good follow-through to give him an accurate call on where my shot broke. Sniper school was unbelievably stressful, and I've often said that I would rather have gone through BUD/S (Basic Underwater Demolition/SEAL training) again than sniper school, because in BUD/S I always felt that if I gave it my all, things would work out. In sniper school, you could throw in everything including the kitchen sink, and just one bad day and you were going home. *Everything* was graded. We started with 24 SEALs in our sniper class, and only 12 graduated. These were seasoned shooters, already wearing the Trident and specifically selected to attend the course because of their professionalism and marksmanship skills. When congratulated later on successful completion of the course and our relative standing as a shooting pair, I always gave the credit to Brandon. The man, as far as I am concerned, was unequaled on the spotting scope and fantastic on the trigger to boot."

—Glen Doherty

**Germans fully geared up to repel a charge**

In Martin Pegler's *Sniper: A History of the U.S. Marksman*, he illustrates the duties of a sniper as outlined in a pamphlet issued to American students:

1. To dominate the enemy snipers, thereby saving the lives of soldiers and causing casualties to the enemy.
2. To hit a small mark at a known range, but without the advantage of a sighting shot.
3. To keep the enemy's line under continual observation and to assist the Intelligence of his unit by accurate and correct reports with map references.
4. To build up and keep in repair his loopholes and major and minor sniping posts.

He also detailed this in more simple fashion as it was put out at the second sniper school:

1. To shake the enemy's morale.
2. To cause him casualties.
3. To stop him working.
4. To retaliate against his snipers.

Photo Credit: Sniping in France by Major H.Hesketh-Prichard

**Effects of light and shade, from Major H.Hesketh-Pritchard's training course**

Simple, to the point, and as relevant today as it was in World War I. With the exception of the Marines, the American forces didn't really have much of an effective marksmanship training program in place, despite sending a few army officers and NCOs to the British sniper schools. The U.S. Marine Corps had been heavily involved in competitive shooting competitions since the turn of the century and because of this, they had continued to push the envelope in training, a tradition that survives today. They distinguished themselves with their Springfield 1903s at the Battle of Belleau Wood, turning back the Germans through effective long-range rifle fire, suffering over 1,000 casualties in the process.

In addition, many Marine snipers were effective on the front with 1903s in combination with the Winchester A5 telescopic sight. The other sight commonly used by army snipers was the Warner & Swasey Model 1913, but the A5 was considered vastly superior, and was still in use as late as World War II.

Another front in World War I that saw the widespread use of snipers was Gallipoli, where Allied forces including the British, French, Australians, and New Zealanders tried to secure a foothold against the Ottoman Turks in an attempt to capture Istanbul and to try to open a supply route to the beleaguered Russians on the Eastern Front.

The trench warfare on the this front included the common usage of periscope sights and jerry-rigged

Lovat Scouts using vegetation to their advantage

LA BRIGADE MARINE AMERICAINE AU BOIS DE BELLEAU


Marine brigade at Belleau Wood

Photo Credit: www.theage.com

Turkish soldiers in a frontline trench at Anzac

"remote" firing systems—scoped rifles on makeshift tripods with strings on their triggers, an Allied attempt to get accurate shots on Turkish shooters without exposing themselves to hostile fire. There are some famous shooters that did battle in Gallipoli, on both sides. The Australian Billy Sing went head-to-head with a famous Turkish sniper

An Aussie sniper uses a "remote controlled" shooting system. His spotter helps key in on targets and calls wind, hits, or corrects for misses.

Rare footage of Gallipoli

nicknamed Abdul the Terrible. The story goes that the sultan sent his foremost sniper after Sing, who had been decimating the Turkish lines (Sing was later credited with as many as 201 kills). After an intense battle over several days, legend has it that the two located each other at the same moment, and Sing placed a shot right between Abdul's eyes a split second before the Turk pulled the trigger. (If you haven't seen the movie *Gallipoli* starring a young Mel Gibson, its worth a viewing . . . great ending.)

The Allies would eventually dominate the sniper battles on the Western Front, with the help of the fresh troops sent over from the United States. Unfortunately, after the war ended, the powers that be, who were reluctant to allow the sniper schools to exist in the first place, ended up dismantling them. There were still

Relaxing in the trenches. Suffering is more easily handled when shared.

some who, despite the success of the Allied snipers, could not help thinking of it as "unsportsmanlike." Truly a shame, as it would not be long before the snipers services would be required again.

Winter warfare brings about a whole new set of problems. Here Soviets prepare to counter an assault in Finland.

## World War II

Most of the world's superpowers had all but dissolved their sniper programs when the German blitzkrieg stormed through Europe in 1939. Russia was the only country that claims to have maintained a full-blown sniper program, although the Germans still had plenty of shooters from World War I. The British had lowered their official snipers to eight per battalion, and in the U.S. forces, only the Marine Corps maintained any sniper training at all, unofficially training just enough scout snipers to keep all knowledge from being lost. There wasn't much place for snipers during the initial invasion through Western Europe, as the reliance on and application of speed, armor, and aircraft kept the battle away from the stagnated lines that were seen 21 years earlier in World War I. Snipers were mostly being used as a rear guard force, slowing down the rapidly advancing Germans. Most people remember Russia as an enemy of Hitler's Reich during World War II, but initially they had teamed with the Germans and invaded Poland together, decisively taking the country and dividing it between themselves.

Shortly thereafter, the Russians would learn the hard way the deadly effect snipers and marksman can have during what would later be known as the Winter War. In the history of war, two soldiers arose from this conflict posting numbers yet to be matched in the tallies of snipers: Simo Häyhä and Suko Kolkka.

The Finnish people had a long history of marksmanship and hunting, and the inexperienced Soviet army of 1.5 million were easy picking for the camouflaged shooters, who operated in familiar terrain, often on skis, in a motivated defense of their homeland. Häyhä is credited with over 500 kills with his Mosin-Nagant model 28 rifle (using open sights), and another 200 using other weapons. He was shot through the face in March 1940, but managed to survive and recover from his wounds, receiving a field promotion from corporal to Lieutenant, the highest jump in ranks in Finnish military history. Later in life he was asked how he became such a good shot. He replied (you have to love this), "Practice."

Photo Credit: Za Rodinu

Simo Häyhä had over 500 confirmed kills, but it is said he most likely had closer to 750.

Suko Kolkka's tallies are also incredibly impressive, and he is mentioned in many books on snipers, but his actual existence is in question. There are no records of him in any Finnish military personnel files, and it is now considered possible that he is strictly legend. With or without Kolkka, the Finns decimated the Soviets, bringing their casualty numbers to the one million mark (while suffering only 25,000 thousand Finnish casualties) and forcing the Soviets to the negotiating table. In the end, a treaty was signed, with the Finns giving up some land and resources to the Russians. Although the following has been quoted in many books on snipers, it is worth repeating: It is alleged that one Soviet general commented after the treaty, "We gained 22,000 thou-

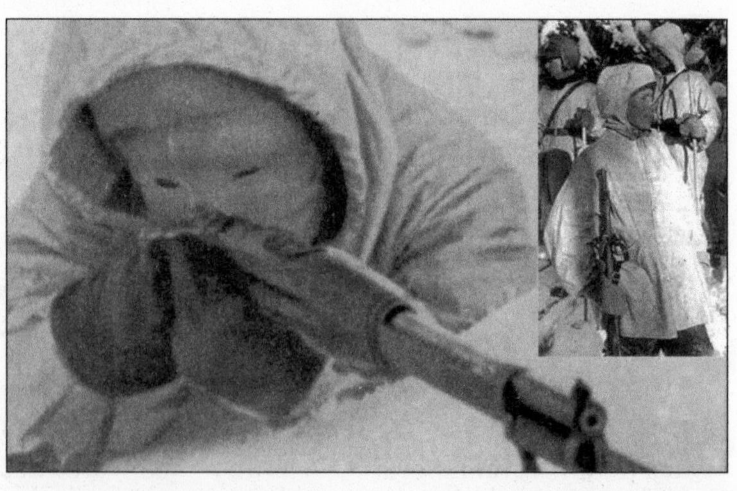

Simo Häyhä knotched an incredible tally of kills using mostly iron sights.

**The relatively green Soviet troops suffered against the Finnish defenders.**

sand square miles of territory, just enough to bury our dead."

Once the war was in full swing, all of the major players got back in the game and started training snipers again. The British opened a school at Bisley, with the help of the famous Lovat Scouts, and provided marksmen with a three-week crash course in sniping: camouflage, observation skills, advanced marksmanship, fieldcraft, and scouting. The British primarily depended on the Lee Enfield No. 4 with a No. 32 scope, which proved to be a great rifle throughout the war and beyond. In the United States, an advanced instructor's rifle course was established at Camp Perry in Ohio, and a sniper school was formed in late 1942 at Fort Bragg, North Carolina, in addition to programs at Camp Lejeune, Camp Pendleton, and Greens Farm in San Diego. Unfortunately, most of the U.S. Army schools focused primarily on marks-

**An example of great winter camo during WWII**

manship and neglected to emphasize many of the other important skills essential to any competent sniper. The other important skills were learned in the field, from other snipers, or in forward training environments like the sniper schools set up in North Africa while the Allies battled Rommel prior to the invasion of Sicily. The U.S. Marine Corps, fortunately, recognized the importance of teaching the "scout" aspect of the title given to scout snipers. In addition to marksmanship, they focused on camouflage, reconnaissance, map reading, directing artillery, and observation. Throughout the war the primary sniper rifle of the Americans was some version of the 1903 Springfield, mostly the M1903A4. With infantry carrying the fantastic M1 Garand, it's a wonder that weapon could not be converted for snipers, but unfortunately a good scope mounting system just couldn't be sorted out until late 1944, so the 1903 became the rifle of choice, accompanied by the 2.5x Weaver 330C scope (mostly used by U.S. Army snipers) or the Unertl 8x scope used by the U.S. Marine Corps.

Under the orders of Heinrich Himmler, the Germans also began a sniper program at Zossen. In Pat Farey and Mark Spicer's *Sniping: An Illustrated History*, they detail the German "Snipers' Code" as follows:

1. Fight fanatically.
2. Shoot calm and contemplated; fast shots lead nowhere; concentrate on the hit.
3. Your greatest opponent is the enemy sniper; outsmart him.
4. Always only fire one shot from your position; if not, you will be discovered.
5. The entrenching tool prolongs your life.
6. Practice in distance judging.
7. Become a master in camouflage and terrain usage.
8. Practice constantly, behind the front and in the homeland, your shooting skills.
9. Never let go of your sniper rifle.
10. Survival is ten times camouflage and one time firing.

The Germans were primarily using the Mauser Kar 98k bolt-action rifle, fitted with a variety of different sights including the Zf41, which reminds me of a lot of modern quick acquisition target sites like the ACOG, Aimpoint, or EOTech. The Germans also experimented with one of the first semiautomatic sniper rifles, the gas-operated Gew43, which was tactically superior in close combat but was not as accurate at range. In

**Barry Pepper playing Private Jackson in *Saving Private Ryan* sighting in with the 1903A4. Okay, it's Hollywood, but he was a great character, and shot lefty to boot.**

1941, the teams that once played nice turned on each other, and Germany invaded Russia.

Here Russian snipers were used to great effect as a rear guard behind the retreating Russian army, taking out officers and NCOs, artillerymen, and lines of communication. The Russians were primarily using the Moisin-Nagant 1891/1930 rifle with scopes made from Zeiss, primarily a 4x scope, but later they mass-produced a 3.5x telescopic sight which was well received. The recordbreaking sniping numbers of the Russian military are extraordinary, with a tremendous list of men and women snipers tallying over 300 confirmed kills. Knowing what we do about the Soviet propaganda machine, it is likely that these numbers are exaggerated, but even if you quartered some of their top snipers' numbers, it is still quite impressive. The Soviets relied on the "cult" of snipers as part of their propaganda program and morale booster for their beleaguered troops, so they chose to idolize them in the press. They were also the first to include women in their snipers ranks, some of whom became legendary. One, named Lyudmila Pavlichenko (the Germans called her the Russian Valkyrie), fought in the defense of Odessa and the

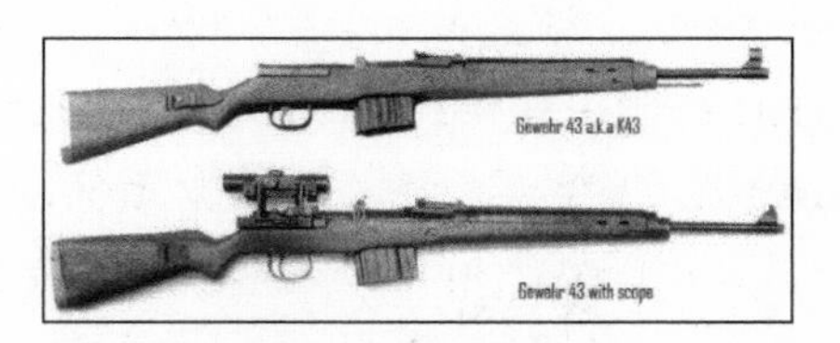


**German Gewehr 43 rifles**

German sniper in WWII

Crimea and was invited by Eleanor Roosevelt to the United States to speak to people about the war on the Eastern Front.

In Andy Dougan's *Through the Crosshairs: A History of Snipers*, he mentions a speech Pavlichenko gave in Chicago. During the speech she said, "I am 22 years old, and I have already destroyed 309 enemy soldiers who have invaded my country. I hope you will not hide behind my back for too long."

The Russian Valkerie

The Soviet sniper we have all heard of, read about, and seen in the Hollywood movie *Enemy at the Gates* (which would have been a great film if they just could have left the love

Lyudmila Pavlichenko during happier days

German sniper in WWII with the Mauser 98K

story out of it—just one guy's opinion) was Vasily Grigoryevich Zaitsev, who was credited with over 200 kills in the defense of Stalingrad alone and was later conferred the decoration of Hero of the Soviet Union. The shepherd from the Urals improvised an academy within the besieged Stalingrad to help train other snipers but is most famous for an alleged duel that took place between a German and himself. If you've seen the movie, then you know how it goes, but whether the Germans sent in their top sniper to deal with Zaitsev or not, from his diaries he did detail a multiple-day stalk where he and two of his shooter/spotters went head-to-head with a German sniper and ended up killing him as he hid beneath a sheet of metal in a contested area.

*Photo Credit: Beevor, Antony* The Fateful Siege, *Penguin U.S.A.*

Vasily Zaitsev teaches new snipers with a Moisin-Nagant and 4x power PE scope.

Zaitsev shows his rifle to a Russian General.

The bloody war on the Eastern Front, what the Russians call the Great Patriotic War, would be the deadliest battleground in the history of the modern world, with an estimated 25 million Russian soldiers and civilians dead and 3 to 5 million Germans killed. Thousands of these deaths were caused by a single shot from a sniper's rifle

Photo Credit: Za Rodinu

Female Russian sniper after being on the losing end of a battle with a German sniper

The Germans also had prolific killers, and once the tides had turned and the Russians were on the attack, German snipers took on the duties of the rear guard. In a book titled *Sniper on the Eastern Front: The Memoirs of Sepp Allerberger,* famous German sniper Allerberger tells author Albrecht Wacker about some of his tactics for stopping an advance:

> I would bide my time until the next four waves were on their way towards our lines, then open up rapid fire into the two rear waves, aiming for the stomach. The unexpected casualties at the rear, and the terrible cries of the most seriously wounded, tended to collapse the rear lines and so disconcert the two leading ranks that the whole attack would begin to falter. At this point I could now concentrate on the two leading waves, dispatching those Soviets closer than 50 meters with a shot to the heart or the head. Enemy soldiers who turned and ran transformed into men screaming with pain with a shot to the kidneys. At this, an attack would frequently disintegrate altogether.

**Running and gunning in Stalingrad**

## D-day

The battle for Western Europe took on a new face with the launching of Operation Overlord, what most of us remember as D-day. The German snipers wreaked havoc on the allied troops, who were fighting a new kind of war in terrain the Normans call *bocage*. Fighting through the hedgerow country of France was hell for the Allies, as German snipers decimated the ranks of officers, NCOs, and medics, to the point where officers began to disguise themselves, carrying the weapons of the enlisted troops, hiding their rank insignia, and keeping their binoculars concealed. On the beaches the corpsmen were targeted, easily distinguishable by the white band with a red cross on their arms; they were often gunned down while dragging a wounded soldier to cover. The thick foliage that bordered the roads and farms was perfect for cloaking snipers, and the German command took full advantage. A single sniper could hold up entire columns for hours or more, and new countersniper tactics had to be brought to bear to keep the front moving toward Germany. The Germans would send out one to three-man teams, with enough food and water for several days. They would operate ahead of and behind the lines at their own discretion, and when supplies ran out, they often just surrendered.

Photo Credit: Robert Capa, gelatin silver print. Cornell Capa International Center of Photography

**American soldier killed by German sniper, Leipzig, Germany, April 18th, 1945**

One can imagine that many a kangaroo court dispensed swift battlefield justice to captured snipers. Snipers were widely reviled and often didn't survive to see the inside of a POW camp. One of the great correspondents of World War II was Ernie Pyle, a Pulitzer Prize–winning war correspondent. He had this to say about sniping in the hedgerow country of France, written in a report filed a few days after D-Day:

> Sniping, as far as I know, is recognized as a legitimate means of warfare. And yet there is something sneaking about it that outrages the American sense of fairness. I had never

sensed this before we landed in in France and began pushing the Germans back. We had had snipers before—in Bezerte and Cassino and lots of other places, but always on a small scale. There in Normandy the Germans went in for sniping in a wholesale manner. There were snipers everywhere; in trees, in buildings, in piles of wreckage, in the grass. But mainly they were in the high, bushy hedgerows that form the fences of all the Norman fields and line every roadside and lane.

It was perfect sniping country. A man could hide himself in the thick fence-row shrubbery with several days' rations and it was like hunting a needle in a haystack to find him. Every mile we advanced there were dozens of snipers left behind us. They picked off our soldiers one by one as they walked down

Photo Credit: Extracted from Medical Service in the European Theatre of Operations

**Infantry duke it out in the hedgerows of France.**

Photo Credit: www.normandybattlefields.com

**A road between two hedgerows near Normandy. It would be pretty hard to pick out a camouflaged shooter in that. Can you see one?**

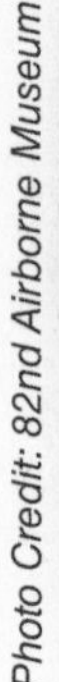

Photo Credit: 82nd Airborne Museum

Soldiers man a machine gun at the base of a hedgerow.

the roads or across the fields. It wasn't safe to move into a new bivouac area until the snipers had been cleaned out. The first bivouac I moved into had shots ringing through it for a full day before all the hidden gunmen were rounded up. It gave me the same spooky feeling that I got on moving into a place I suspected of being sown with mines.

In past campaigns our soldiers would talk about the occasional snipers with contempt and disgust. But in France sniping became more important, and taking precautions against it was something we had to learn and learn fast. One officer friend of mine said, "Individual soldiers have become sniper-wise before, but now we're sniper conscious as whole units."

Snipers killed as many Americans as they could, and when their food and ammunition ran out they surrendered. Our men felt that wasn't quite ethical. The average American soldier had little feeling against the average German soldier who fought an open fight and lost. But his feel-

Ernie Pyle was killed in the Pacific Theater on the small island of Ie Shima off Okinawa. Here he is buried along with other soldiers killed during the assault.

German sniper in Normandy

ings about the sneaking snipers can't be put into print. He was learning how to kill the snipers before the time came for them to surrender.

Marines check out dead Japanese snipers off the island of Leyte Phillipines.

## Pacific Theater

Both the U.S. Army and Marine Corps saw intense combat in the Pacific, with the Marines doing most of the island hopping and the Army seeing its action in the Phillipines and on the Southeast Asia mainland. On many of the islands taken, it was difficult to employ the sniper with traditional equipment, with thick jungle fighting limiting the range to the fanatical Japanese defenders. Other islands like Okinawa, Saipan, and Iwo Jima were much more open, and there are plenty of recorded stories of U.S. Marine snipers taking out the enemy at ranges greater than 1,000 yards.

Marine snipers in Okinawa scan for Japanese soldiers with their 1903s and Unertl scopes. Allegedly, shortly after this photo was taken, one of the shooters dropped a Japanese soldier at 1000 yards.

*Photo Credit:* Life's Picture History of World War II., *Time Incorporated New York*

**Beach assault in Saipan**

The Japanese snipers were primarily using the Ariska Type 97 sniper rifle, which shot a 6.5mm round and used a 2.5x telescopic sight. Later in the war, a shorter-barreled rifle was released, the Ariska Type 99, which transitioned to a larger 7.7mm round and had a more powerful scope on it as well. The Japanese snipers followed the Bushido code, and were honor-bound to accept death before surrendering in the name of their emperor and homeland. In the tight jungle warfare, they were often found high in palm trees, well camouflaged and waiting for the troops below to pass before targeting officers and NCOs from behind to create more havoc. Iron sights and machine guns were particularly useful in the close quarters of jungle warfare: U.S. forces would handle the Japanese snipers more often than not by spraying the treetops with the Browning automatic rifle (BAR) or Thompson machine guns. If they were taking fire from farther in front, with a clear field of fire, then countersnipers could be used, often taking out the enemy machine gun nests with accurate fire from over a kilometer out. Later, toward the end of the war, it wasn't uncommon for the Allies to just call in the heavy artillery and completely level entire areas to clear out one sniper. Necessity is the mother of invention; nothing was out of the realm of possibility. Dogs were used, grenades, satchel charges, howitzers, antitank recoilless rifles—you name it. But, when in doubt, blast them out.

The Japanese may not have been the best marksmen, but they were renowned for their skills at camouflage and jungle survival. Farey and Spicer's *Sniping: An Illustrated History* quotes a 1943 *Time* magazine article:

> Marine and Army men returning from the South Pacific almost unanimously hold that, man for man, the Jap soldier is inferior in fighting qualities to the American. But in all the things to do with hiding, stealth, and trickery, they give the Japs plenty of angry credit. They dig deep, stand-up foxholes, which are safe except

Photo Credit: National Archives

**Marines advance with the Thompson machine gun and the BAR, both great weapons used to deadly effect countering Japanese snipers.**

U.S. troops check out a Japanese snipers' nest on Papua New Guinea.

under direct artillery fire (and which are better than U.S. slit trenches). On the defensive, they dig themselves dugouts protected by palm trunks, and then they crawl in and resist until some explosive or a human terrier kills them. The myth of the Japanese sniper is exploded by returning officers. They say that Japanese snipers are an annoyance, little more. They hide excellently but their aim is poor. Sniping serves, however, to frighten men who will not deliberately ignore it. . . . But the greatest handicap of the Japanese is their lack of imagination. They carry out orders to the letter and, if necessary, to death. But when things go wrong, they cannot adapt their tactics. If Jap attackers meet resistance, they advance anyhow—which accounts for the terrible slaughter to which Japanese troops submit themselves.

The terrible slaughter of the Japanese had yet to reach its zenith, and it would not be the sniper that would bring an end to the war in the Pacific. On August 6, 1945, a B-29 named the *Enola Gay* would drop the first atomic bomb, named Little Boy, on Hiroshima. Three days later, a second bomb would level Nagasaki, and shortly thereafter World War II would come to an end.

## Korea

The peace was tenuous in the years following World War II as the world witnessed the arrival of two superpowers that would spend the next 40 years duking it out over the hearts, minds, and bodies of the world's population. Lines were drawn and the Western allies geared up to stem the tide of Communism. In the new Atomic Age, little thought was given (again) to the relevancy of the lone shooter, and the allies continued to thumb their noses at history and allowed their sniper programs to go the way of the dodo bird. Not so much the Communists. The Russians had witnessed and valued the destruction their shooters inflicted on the German ranks and passed on their tactics to other Communist states. In June of 1950, the fragile peace that existed between Communist North Korea and Democratic South Korea dissipated as the North pushed below the 38th parallel, sparking a bloody, three-year war. Initially, the Chicom (Chinese and North Korean) soldiers wreaked havoc on the Allies with accurate sniper fire from primarily World War II–era Moisin-Nagant sniper rifles with PU or PE scopes.

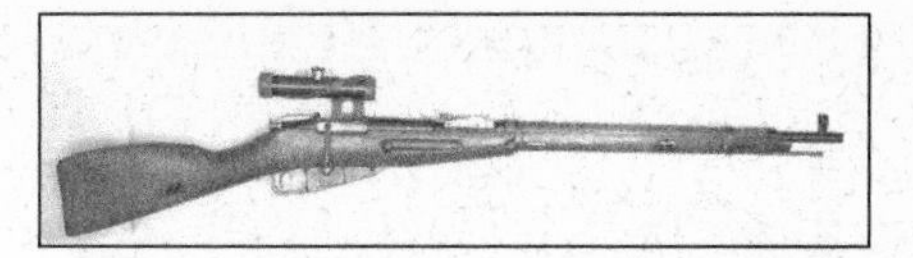

**Moison-Negant sniper rifle with PU scope**

As usual, despite pressure from the actual combatants, a senior officer had to be personally affected before anything actually got done. One of the stories from the war tells the real start of the U.S. in-house sniper program in Korea. In Haskew's *The Sniper at War*, the classic story from Adrian Gilbert is retold. The commander of the 3rd Battalion, 1st Marines allegedly was trying to survey the field one morning when, in Haskew's words:

> He placed his binoculars in the bunker opening and gazed out. Ping! A sniper bullet smashed the binoculars to the deck while blood welled up in the crease in his hand. The battalion commander, fortunately, was only scratched, but he reflected that it was a helluva situation when the CO could not even take a look at the ground he was defending without getting shot at. Right then and there he decided that something had to be done about the enemy sniper. The colonel learned that in the supply section there were an adequate number of rifles and

telescopic sights. He next sent for an experienced gunnery sergeant who had spent time firing with rifle teams. He told the gunny what he wanted. The gunny visited each company to pick sniper candidates. He wanted riflemen who possessed the characteristics of good infantrymen. But above all, he stressed the need for patience. This trait is absolutely essential, for a sniper must remain still and alert for long hours, waiting for the enemy to show himself. Soon the range was ready, and the gunny began an intensive three week course on sniping . . . after training, the Marine teams were sent out to the various outposts. To spur them on, a case of cold beer was awarded to the men of each outpost that got 12 kills within a week. All hands turned to helping the rifle experts in spotting enemy snipers. Only a week after the sniper teams went into action, the division commander came to test their efficiency. The two-star General strode the entire length of I company's Main Line of Resistance armed with nothing but his walking stick. . . .

*Photo Credit: Rickard, J (17 November 2008),"Japanese Snipers' Nest," New Guinea*

**Chinese sniper Zhang Taofang allegedly had 214 kills in just over a month using an old Moisin-Nagant with iron sights.**

Photo Credit: Department of Defense Visualization Center

**USMC sniper pair working the field in Korea**

The Americans had to dig into the storage units to get outfitted for Korea and went back to using the M1C Garand semiautomatic or the Springfield 1903 bolt action.

In close combat, it was invaluable to have a quick follow on shot, but over longer ranges, the 1903 was still preferred. The M1 still suffered from two major drawbacks: first and foremost it was still loaded by a metal clip that would eject when the last round was fired with a loud "clang" . . . not so good when trying to conceal your location; also the enemy would know you were reloading. Secondly, the optics still had to be mounted off to the side, which didn't make for ideal body position and eye relief. Modifications were becoming available though; add-ons in the form of leather pads for the stock for more appropriate cheek welds and flash suppressors were becoming readily available to full-time shooters. In-country armorers also undertook several interesting experiments. Modifications were made to the Browning automatic rifle, such as mounting a Unertl 8x scope and firing the weapon in single-shot mode off a bipod or tripod, which proved accurate to

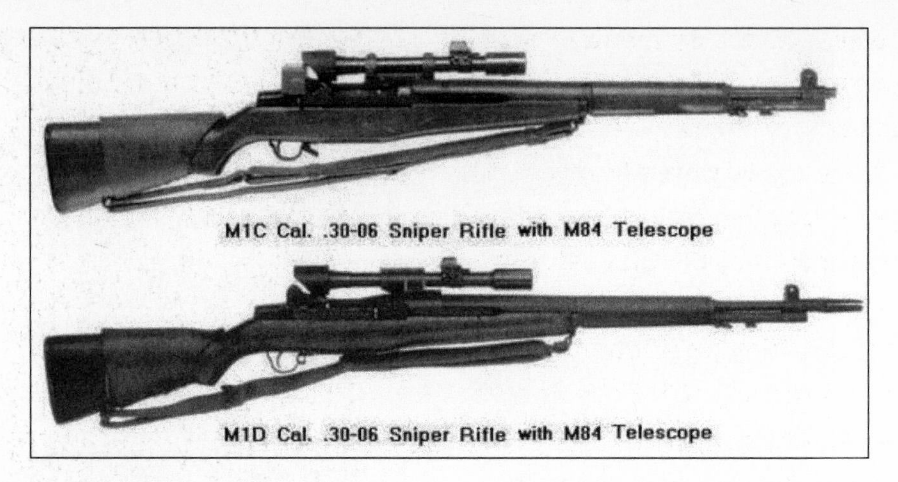

M1C Cal. .30-06 Sniper Rifle with M84 Telescope

M1D Cal. .30-06 Sniper Rifle with M84 Telescope

2,000 yards. Other talented inventors played with captured Soviet antitank rifle actions and Browning barrels to create various .50-caliber long-range sniper weapons. In *One Shot One Kill*, Charles Sasser and Craig Roberts tell another classic story of sniping on the front lines during the Korean War. U.S. Army Corporal Chet Hamilton was watching the assault of a steep hill (near the famous Pork Chop Hill) and was in position to see his boys take a beating from the Chinese Communist troops, known at the time as "Chicoms":

> I felt helpless watching from the sandbagged trenches . . . until I noticed something. It was only about 400 yards across the valley from the Chinese lines. My position put me on almost the same level with the chink defenders on the other hill. In order for the Chicoms to see our troops and fire at them down through their wire as the GIs charged up the hill, they had to lean up and out over their trenches, exposing wide patches of their quilted hides. (Chicoms often wore quilted uniforms for protection during the ridiculously cold winter months. The Allies weren't so fortunate.) That was all I needed. It had become a clear morning in spite of the smoke and dust boiling above the Chinese hill. The four-power magnification of my scope made the chinks leap right into my face. All I had to do was go down the trench line, settle the post-and-horizontal-line reticle on one target right after the other, and squeeze the trigger. It was a lot like going to a carnival and shooting those little toy crows off the fence. Bap! The crow disappeared and you moved over to

> the next crows. By the time you got to the end of the fence, you came back to the beginning and the crows were all lined up again ready for you to start over. I don't know who the Chinese first sergeant was over there, but he kept throwing up another crow for me every minute or two. And I kept knocking them off the fence. The fight for the hill lasted about two hours. The other guys . . . came to watch, point out targets, and cheer when I zapped one.

The shortsightedness of U.S. military commanders continued after the signing of a peace treaty in 1953, and the art of the seasoned military sniper was once more shelved in the practicality closet and replaced with the idea of global thermonuclear war. Some advocates, notably Warrant Charles Terry and Lieutenant Jim Land who were shooting on the U.S. Marine Corps team, fought to keep the hard-earned skill sets alive and well in the U.S. military. Those skills would be needed again soon.

## Vietnam

Leave it to the careerists and politicians to ruin a perfectly good thing. I realize I've been beating a dead horse, reiterating the fact that in every major conflict the United States was involved in, the sharpshooters and snipers proved their value time and time again, only to have politics and weak leadership inevitably intervene and squash whatever programs were in place, pushing the true battlefield commanders behind desks, and letting the rest of the talent go where they may. Vietnam wasn't much different.

**Signed photo of the legendary Carlos Hathcock**

The U.S. Marines were the first conventional troops to enter Vietnam in March 1965, and they quickly came under accurate and constant fire from seasoned Viet Cong (VC) guerilla fighters and North Vietnamese army regulars who had been fighting for independence for over a decade. This was jungle warfare all over again, and throughout the course of the war there were very few large-scale engagements with two armies in a pitched battle toe-to-toe. It was guerilla fighting, hit and run, ambushes, booby traps, and highly trained snipers.

Captured instruction manuals give us a look into what was a well-run and intensive North Vietnamese sniper course, most likely aided by Soviet advisers. In addition to teaching long-range shooting, they taught basic armorer skills so that shooters could repair their weapons in the field. They were also taught camouflage, observation, and reconnaissance and would take these skills south to teach to their guerilla counterparts—the VC. A great example of the effectiveness of one well-trained sniper is given in Adrian Gilbert's book *Sniper*:

> One such instance was observed by a Marine sniper, Joseph T. Ward, who described how a communist sniper had pinned down a neighboring company of Marines. Although an air strike had been called, low clouds prevented the three-strong flight of F-4 phantoms from attacking for over two hours. During this time, whenever a Marine moved he was picked off by the sniper. Even after the first strike of napalm, the sniper continued his work, wounding another man before departing the scene as the second strike went in. Obliged to admire the skill of the persistent sniper, Ward recalled: "While we cleaned our rifles, I thought what a day's worth one of Uncle Ho's best had given. One enemy sniper had killed three grunts, wounded four, tied up two companies and six fighter-bombers a good part of the day, and we hadn't seen any sign of him."

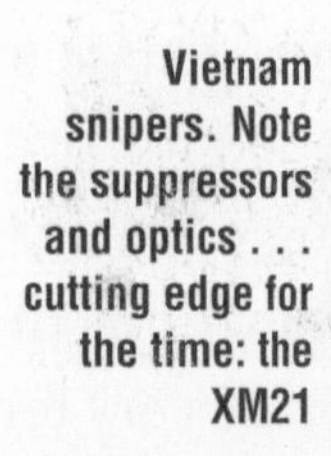

Vietnam snipers. Note the suppressors and optics . . . cutting edge for the time: the XM21

With no official scout sniper program in place (there was a small program being run in Hawaii in the early 1960s by Captain Jim Land), a program was quickly set up in country, recruiting veterans of the U.S. Marine shooting team as instructors. With pressure from above, Captain Bob Russell and later Captain Land each set up Marine sniper schools within the relative vicinity of Da Nang, bulldozing thousand-yard

Vietnam era XM21 sniper system

ranges into the bush and creating their own curriculum from lessons they learned by personally taking out VC in the field. Land was under strict orders from the commander of the 1st Marine Division: "I want you to organize a sniper unit within the First Division. Captain Russell in the third division started training snipers last year. I want mine to be the best in the Marine Corps. I want them killing VC and I don't care how they do it, even if you have to go out there and do it yourself." One of Land's first students was a bored former MP who would later become a sniping legend. That man was none other than the White Feather (Long Tra'ng, in Vietnamese, because he wore a white feather in his hat), Carlos Hathcock himself. Already a decorated competition shooter, Hathcock was a natural for the job and has gone down in history as having some of the most impressive stalks, kills, and survival stories to come out of the war. Some of his shooting highlights include

Photo Credit: The National Museum of the Marine Corps

Painting of a sniper in Vietnam

93 confirmed kills and a shot with a Browning .50 caliber modified with a scope that dropped a VC at 2,500 yards. There were snipers with more confirmed kills (U.S. Marine Chuck Mahwinney: 98, U.S. Army sniper Adelbert Waldren: 109), but none were as well-known.

His tale has been told so many times, suffice to say I just want to praise him as a true hero, and relate what I deem his most heroic act, which didn't even involve a rifle. While traveling in a half-track with some other Marines, the convoy was shelled and his vehicle burst into flames. Despite being on fire, he helped rescue all the other Marines in the truck, before barely escaping himself, with third-degree burns covering 40 percent of his body. It took over a year for him to recover, and for this heroic act he was awarded the Silver Star. It also effectively ended his active sniping career, but his knowledge and experience wasn't allowed to go to waste (for once!), and he later became an instructor at the U.S. Marine Corps sniper school which opened in Quantico, Virginia, in 1977.

The Marine Corps' push for new snipers also eventually brought about a new weapon, the Remington 700, soon to be known as the M40 sniper rifle, with the Redfield Accurange 3-9x scope replacing the old Winchester Model 70. (The Remington is a fantastic weapon and still in use today). The U.S. Army, typically bringing up the rear (kidding Army brothers, kidding), didn't really fully adopt a sniper program until June of 1968 when they brought members of the AMTU

Photo Credit: Pfc Charles B. Mawhinney by Michal W. Wooten © 2003 Military Art Gallery

Chuck Mahwinney portrait

(Army Marksmanship Training Unit) from Fort Benning, Georgia, to Vietnam. They showed up with a few leftover sniper rifles, some old Garand M1C and M1Ds, along with the classic Springfield M1903 A4 that was paired with the incredibly average 2.2x M84 telescopic sight, but established a difficult and intensive 18-day training program. Back in the states there was a push to find a new weapon. After extensive testing at home, the Army settled on what would become the XM21, a match grade M14 fitted with the Redfield 3-9 power scope with some cool new technology called ART (Automatic Ranging System). This system essentially took out the need to range your target inside of 600 yards, by automatically adjusting for the range so long as the shooter put the targets body in between two lines in appropriate fashion using a special "cam" mechanism. Other technology, including new night vision optics and suppressors for the rifles, proved to be extremely effective in the field—this is where the United States truly began to "own the night." It took a while, but the Army finally realized what the Marines so aptly summed up on a sign on the outside of their sniper school in Quantico, Virgina: "The average rounds expended per kill in Vietnam with the M-16 was 50,000. Snipers averaged 1.3 rounds. The cost difference was $2300 vs. 27 cents." Despite producing highly qualified snipers in country, the big Green Machine still couldn't figure out initially how to employ them. In Martin Pegler's *Sniper: A History of the U.S. Marksman*, a U.S. Army private sums up the situation perfectly:

Photo Credit: www.olive-drab.com

**Sniper Adelbert Waldron, Vietnam 1969**

> Although I trained as a sniper I was just another dog faced grunt in the squad, pulling patrols and guard. I didn't even get issued a sniper rifle, only the M16. When we got hit by a VC sniper the

lieutenant started yelling for me, but with the M16 I couldn't do anything. I went to the captain and told him sending me on a sniper course then using me as a grunt was stupid. He agreed and I got an immediate issue of an XM-21, which really pissed the lieutenant off. He kept putting me on point and it was some weeks before they started pairing us up and really using us for what we'd been trained for.

Once they did get properly employed, they were incredibly effective, as they proved in 1969, where from January to July they attributed approximately 1,139 confirmed kills. Cost effective, deliberate, low maintenance, and deadly: That was the U.S. sniper in Vietnam.

This is Lance Corporal Dalton Gunderson, USMC, showing a modified seated shooting position.

Hathcock was credited with hitting a NVA at over 2,000 yards with a .50 cal mounted with a telescopic scope and set to fire on single shot.

## To the Present

Since Vietnam, the United States has been involved in several conflicts while policing the world: Beirut, Grenada, Panama, Somalia, Bosnia, Afghanistan, and Iraq to name a few. The U.S. sniper played a part in all of these, and there are great stories that have come out of each conflict. On a downside, the professional sniper-soldier has a new and ever growing problem: politically correct war fighting. Since becoming the world's police force, engagements are consistently bogged down with unrealistic ROEs (rules of engagement) and policy making from people thousands of miles away who aren't on the front lines. In Beirut in the early 1980s, there were several documented incidents of Marine snipers being forced to watch atrocities and not being autho-

Randy Shugart and Gary Gordon

A view not many get to see, but those that do don't ever forget. Mil Dot scope reticle.

rized to engage. One sniper watched as "a whole company of unarmed Lebanese soldiers was massacred in cold blood." Pegler writes:

> Another team coming under fire in Beirut in 1983 was denied permission to return fire, as no one could find an officer senior enough to give his authority. Fortunately other NATO peacekeeping troops serving alongside the Americans did not require such a convoluted system and simply shot back, usually to great effect. One frustrated Marine sniper commented that "We aren't keeping the peace, the Israelis are—they got loaded weapons. The ragheads know they'll get their asses shot off if they pull anything with them."

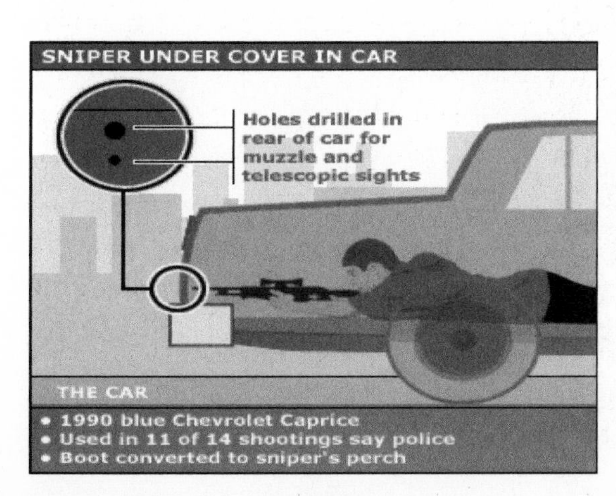

SNIPER UNDER COVER IN CAR
Holes drilled in rear of car for muzzle and telescopic sights
THE CAR
• 1990 blue Chevrolet Caprice
• Used in 11 of 14 shootings say police
• Boot converted to sniper's perch


In Somalia during Operation Continue Hope, two Delta Force snipers who were posthumously awarded the Congressional Medal of Honor were immortalized for their sacrifice in the book and movie *Black Hawk Down*. Master Sergeant Gary Gordon and Sergeant First Class Randy

Photo courtesy of Scott Tyler

Scott Tyler on walkabout with the locals in Ramadi

Shughart volunteered to be dropped into a very hot area to secure the crash site. After pulling the only surviving crew member, pilot Michael Durant, to safety, the two men killed countless Somalis before finally succumbing to the endless tide of militants pouring into the location. Durant survived 11 days in captivity, and is alive today because of Gordon and Shugart.

There are some notable shootings that many civilians in the United States would consider "sniper" attacks: Lee Harvey Oswald shooting John F. Kennedy in 1963; Charles Whitman in the University of Texas tower in 1966; John Allen Williams Muhammad and John Lee Malvo, the "D.C. Snipers" in 2002. All of these men, except Malvo, were former Marines, and all committed murder. To call them snipers is to discredit the men and women who performed as one in combat; to call them murderers or psychopaths or terrorists is much more appropriate. I don't want to waste any more time or words on them.

Some artwork and writing from Malvo in prison

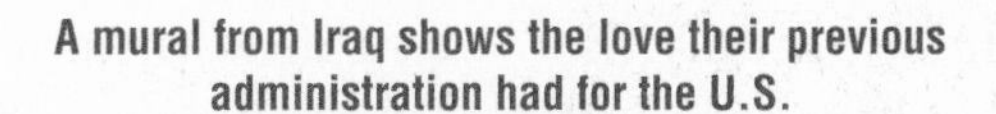

**A mural from Iraq shows the love their previous administration had for the U.S.**

One of our close friends and an infamous sniper in the teams tells a story from when he was in Somalia. "I was on the beach and cleared hot on multiple enemy targets and the area was crawling with them. My main area was almost a 1000 yards away. I had no real help spotting and both the wind and mirage were up. I had two targets over 800 yards away, one skinny with an RPG and the other acting as his spotter and security. I could see the RPG guy getting ready to fire, so I figured I'd take him out first. Maybe only one or two guys in the world could have made that shot in those conditions. I was not one of them! I didn't hold for enough wind and when I fired I saw the guy next to him explode from the impact of my .50-cal round into his top chest. The guy was literally blown to pieces, chucking flesh backwards 10, 20 yards. The skinny with the RPG was so completely freaked that his buddy just exploded that he dropped his RPG and made a run for it. I imagine he's probably still running. Ha-ha . . ."

—Lt. Scott Tyler, former U.S. Navy SEAL

The weapon actually used by the D. C. Snipers was a Bushmaster AR with iron sights. The hide site created by the pair was actually quite ingenious, and has been copied by many snipers overseas. In what was the largest manhunt in history, these two men practically shut down an entire region. They demonstrated the raw power and effect a sniper can have on a community, destroying morale, interrupting all routine, and instilling fear.

**Ramadi Iraq 2005**

Being out at night was always the best time to set up positions and wait for the insurgents to come out in the morning hours to set up their ambushes and IED's (Improvised Explosive Devices). They seemed to like to gravitate toward the same positions over and over again. The spots they chose were usually fairly tactically advantageous. The insurgents who chose poorly oftentimes didn't get a second chance to improve their positions. The problem with the insurgents is that although they were able to pick decent positions to set up on the U.S. Military units in the area, the fact that they would use the same positions over and over created a pattern for us to try and exploit. We were a small number of snipers trying to cover a huge area of city in a place where the cost of failure was high and the rewards were a hot meal and knowing that there were a few less terrorists in the world.

When my Team first arrived in country, we split up and took different locations throughout the country of Iraq. My task unit headed west and took responsibility for the Al Anbar province, which included Ramadi and Fallujah. Fallujah was the city where the four Blackwater employees were ambushed, mutilated, burned, and hung from the overhanging beams on a bridge at the cities perimeter. By the end of 2004, the city of Fallujah had the crap kicked out of it by the Marines in an operation called Vigilant Resolve. It was

Photo courtesy of Scott Tyler

**Scott Tyler enjoys one last cup of coffee before going into the field.**

one of the most dangerous cities in Iraq up until that point, completely lawless and out of control. It was as close to an operational safe haven as possible for local insurgents and Al Qaeda in Iraq (AQI). After Fallujah fell and the terrorists were killed or driven out of Fallujah, the activity seemed to pick up in Ramadi, which was the next city over to the west. When my unit was out there the Marines and Army were losing several people a week and sometime as many as five or more. This is a significant number especially considering that an average Marine Battalion is only around 500 people, give or take a hundred, and not all of them are out beating the streets on patrols.

The best way to understand the situation on the ground in an area where you are working is to walk it. This is exactly what me and the rest of my sniper detachment would do. We would embed with the Marines or with the Army, whichever unit had the operational control of the area (AO). Usually once we arrived in the area we were going to operate in, one or more of us would head over and talk to the commanding or operations officer who would give us permission to liaise with their senior enlisted non-commissioned officers, the guys who ran the troops. From that point we would plan and coordinate patrols, link ups, break offs, pick ups, and contingency actions. Another important step in the process was identifying hot spots, areas where insurgents tended to attack repeatedly. From this information we would start conducting map studies to find areas that we could get visibility on these locations and had a good chance of having a clear shot. After the preliminary analysis we would attach to the conventional units and patrol with them during the day, to get firsthand looks at the potential hide sites, different look angles, and firing positions. I would usually take several photos from different angles and perspectives and then mark it on my GPS so we could return later and set up shop.

There were strict curfews imposed in Al Anbar during the nighttime hours and the insurgents knew very well that if they were caught out at night, when most of the citizens were inside, the odds that they would be killed or captured was great, especially if they were caught in the act of setting up an IED or ambush. This meant that in order for them to be effective and increase their chances of survival, they needed to blend in with the local population and try and carry out their nefarious plans while remaining unnoticed by coalition forces. This situation made nighttime the best for us to move in to get to a location and set up a hide site or firing position. But once in position, it was usually a long wait till daylight when our teams would be on hyper alert, scanning and watching the crowds for anyone doing anything suspicious. Once someone flagged themselves, the burden was on us to positively identify the perpetrator and intended actions and then wreak immediate and final judgment.

**Sniper School 2002**

My most fond memories of sniper school include an introduction by the SEAL Lead Petty Officer instructor who was proctoring our class. As I sat in my seat what I heard was, "If you don't lay in bed at night, hoping and praying that someone breaks into your house so that you can shoot them and kill them, you don't belong in this class." To folks who don't have any exposure to the tradition of arms, this may seem like a horrific and offensive approach to something as sacred as taking the life of another human being. The reality of the situation is that there are monsters in the world who prey upon others, who will take as much as they can unless someone is willing to step forward and stop them. There are also people who live in fear of those monsters and then there are the people who embrace the potential reality of what it means to have to stand up and take the fight directly to those who terrorize others. I would say that I fall squarely into the third group. The actual between-the-lines interpretation of what was said is more akin to wanting to be in the situation where you are forced to take action. No one I worked with just wanted to go out and kill people because they could; we were all motivated to be the people trusted with the skill, intuition, and judgment to make the right calls at the right times. When you are in a hide site, isolated from any support, in the middle of bad-guy land and all you have is your gun and your shooting partner, you want to be next to someone who is not horrified at the prospect of having to take a life or lives when the moment comes, because it will come, and your survival depends on unflinching determination and resolve, enabled by training, to eliminate the enemy before they eliminate you.

—Lt. Scott Tyler, former U.S. Navy SEAL

2

# The 21st-century Sniper Defined

*Refer to the color photo insert at the center of the book when reading this chapter.*

In Vietnam, it took the big military machine an average of 2,000 shots to achieve one confirmed North Vietnamese army kill. It took a U.S. military sniper an average of 1.3. Talk about your cost-effective war machine. After the evacuation of Hanoi, the United States entered a relative lull in global hostilities, taking part in only a few small conflicts over the course of the next 16 years prior to the September 11, 2001, terrorist attacks. The loss of the experienced war fighter brought about a bureaucratic military establishment that eventually would forget the professionalism and power of one well-trained marksman. The 21st century has brought about a shift in training and tactics whose effectiveness is proved immediately in real-time combat. While it would be irresponsible for us to go into detail about modern sniper training methods, it is important for us as former SEAL snipers to give the reader a brief outline of what a SEAL sniper student is up against. The U.S. Navy SEAL course is divided into three phases over 90 days, and it tests to the highest standards in the world.

In phase one the candidate learns the latest in digital photography techniques, computer image manipulation/

Snipers aren't always on the gun. Here a student sets up his camera for some long-range photographs.

compression, and satellite radio communications. Historically the sniper would sketch a target in detail and record notes with pencil and paper. In the 21st century, the sniper leverages technology to his advantage and uses the most advanced camera systems, hardware, and software available to record target information and produce a firing solution.

Phase two is the scout portion of training. The name of the game is stealth and concealment. In this phase the sniper learns the

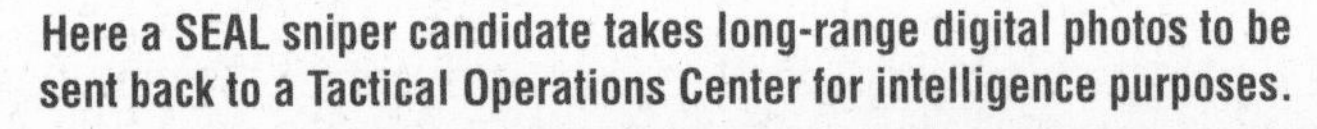

Here a SEAL sniper candidate takes long-range digital photos to be sent back to a Tactical Operations Center for intelligence purposes.

A SEAL sniper candidate veges up his tripod and camera to avoid detection yet still get the intel.

Never underestimate the sniper's importance to reconnaissance and surveillance.

Instructors brief snipers on the "rules" prior to beginning a stalk.

Not always warm and dry

A student tries to locate the OP (observation point) without getting spotted.

Stay low and use the dead space created by the tree to avoid detection.

Find the shooter . . .

Good camouflage

**Instructors man the OP table, using high-powered optics to try to spot the students . . . and fail them.**

art of camouflage, small unit tactics, patrolling techniques, and most important, how to get in and out of hostile enemy territory undetected, without leaving a trace. We often fail candidates who leave behind the smallest trace; a bullet casing left behind will get you sent home. Toward the end of this phase we introduce advanced marksmanship fundamentals and a system of mental management used by the world's top athletes. Mental management gives the students the tools (whether or not they use them is up to them) to cope with adversity and also a system to rehearse and practice their skills perfectly through mental visualization techniques. To prove the value of mental management and rehearsal, I would often relate a true story related to the topic. A Navy fighter pilot was shot down in Vietnam, captured, and imprisoned for years in the famous prisoner of war camp, the Hanoi Hilton. The pilot was an avid golfer back home and to get through the extremely demanding situation he would shoot rounds of golf in his head. For years he would play his favorite courses perfectly in his mind. Eventually liberated and back on U.S. soil, the first thing this pilot did was jump out of the military ambulance and onto the golf course. After explaining his ragged looks (he was a tall man and extremely skinny from malnutrition), he shot nine holes of golf at one under par. This was shocking to those who witnessed the event,

**Rain or shine, sniper candidates spend hundreds of hours on the range.**

and when questioned about how this was possible, the pilot replied, "Gentlemen, I haven't hit a bad shot in four years!"

Phase three is the sniper portion. We spend hours in the classroom learning the science behind the shot, ballistics, environmental factors, and human factors, and calculating for wind, distance, and target lead. We then put the knowledge to practical application on the shooting range. The students train and test with moving and pop-up targets in high-wind conditions out to 1,000 meters. As part of the training, we put the shooters in the most stressful and challenging situations imaginable. We look for signs of high intelligence, patience, and mental maturity. Then we intentionally (often unknown to the candidate) place the shooter in adverse and unfair situations to test their mettle. An example of this would be the "edge" shot. Individual trainees are lined up on the shooting range and are told that they have 4 minutes to run 600 meters, set up on the firing line, and wait for their targets to appear sometime between 4 minutes and 1 second, and an hour. We always send a target up right at 3 minutes—usually right when the shooters are just getting set up on their lanes and identifying their fields of fire. Often, a shooter will take his eyes away for a split second to wipe sweat from his brow, then drop down on his scope to see his target disappear and his opportunity gone. The peer pressure is intense and shooters often break down in frustration at a missed shot. They eventually learn to control their feelings or they don't move on. As instructors we keep detailed student records and record everything. A large percentage of SEAL candidates don't make it through the course and just getting a billet is extremely competitive. No one wants to go back to their SEAL team deemed a loser for having failed out of the course. However, this course is one of the few courses you can fail as a SEAL and not be looked down upon by your teammates. This is because the SEAL course is renowned for being one of the toughest and most challenging courses in the world. More than 3 months of 7-day, 100-hour workweeks go into the training. It takes extreme perseverance in order to graduate with the title of SEAL sniper. It remains one of the most stressful events of my life—even when compared to my combat tours.

The training that takes place in modern military-grade sniper communities is far more advanced than anything seen throughout history.

A major shift in training methodology is the fact that modern-day Special Operations snipers are trained and deployed as independent shooters, not in the traditional

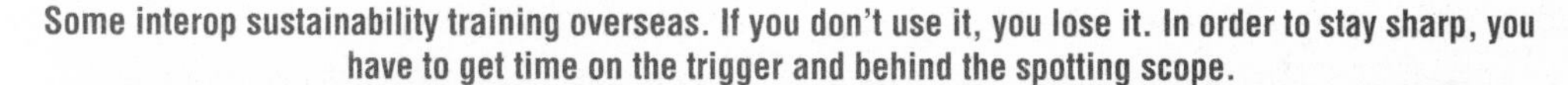

**Some interop sustainability training overseas. If you don't use it, you lose it. In order to stay sharp, you have to get time on the trigger and behind the spotting scope.**

SEAL sniper student well-concealed and ready to shoot!

shooter/spotter pairs. Training in the past would have graduates that were great shots but not so great behind the spotting scope and vice versa. As tough as it is to admit, we were graduating students that were not well-rounded snipers. Because students were trained in pairs and shared scores, often they would make up for each others' deficiencies, and this is not a good way to conduct business or create a world-class sniper. Ninety percent of our sniper missions in the SEAL teams are conducted with single shooters in separate positions, and when I became the course manager, I pushed our officer in charge to accept that if most of our snipers were being used independently, then we needed to start "training like we fight" and focus on graduating a well-rounded sniper capable of operating on his own without a spotter. My thinking was that the first time a sniper operates independently should not be on a real world mission, it should happen in training. We did start to focus on gradually shifting training and testing to ensure that our SEAL students had a complete grasp of all the subject matter. My prediction is that modern courses will continue to shift toward the use of more technology and graduating snipers capable of deploying as single shooters who don't require or rely on a spotter. The modern graduate can deploy in any combat theater without the aid of a spotter, because he has the training and a

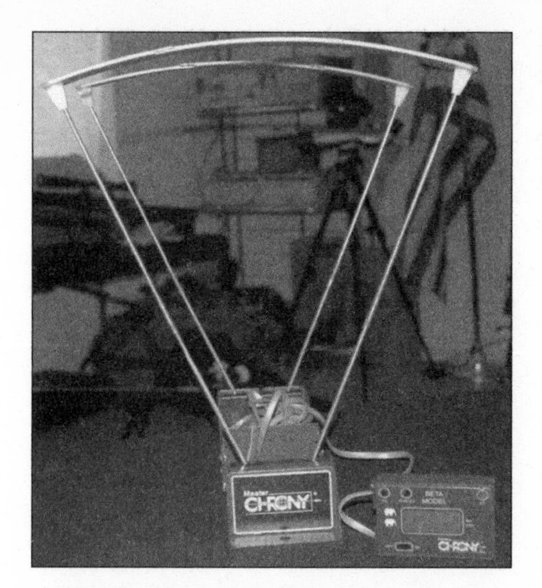


A Shooting Chrony Beta Master Chronograph. It will accurately measure the muzzle velocity for a particular round and rifle combination. That data will then produce an unbelievably accurate first-round.

complete suite of technology at his fingertips. The days of dope books and hand sketches are falling to the wayside and being replaced by digital imagery, handheld computers with complex ballistic software programs, chronographs that measure each weapon's specific muzzle velocity (two identical rifles shooting the same round will produce different results), nanotechnology applied to camouflage, and extremely accurate rapid-fire smart weapons.

The 21st-century sniper is a mature, intelligent shooter that leverages technology to his deadly advantage. He has spent thousands of hours honing his skills. He is a master of concealment in all environments, from the mountains of Afghanistan to the crowded streets of Iraq. He is trained in science, but he alone is left to create the individual art of the kill. The battlefield to the sniper is like a painter's blank canvas. It is up to him to use his tools, training, and creativity to determine how that final shot will play out and the devastating psychological impact that is ultimately the result of his actions.

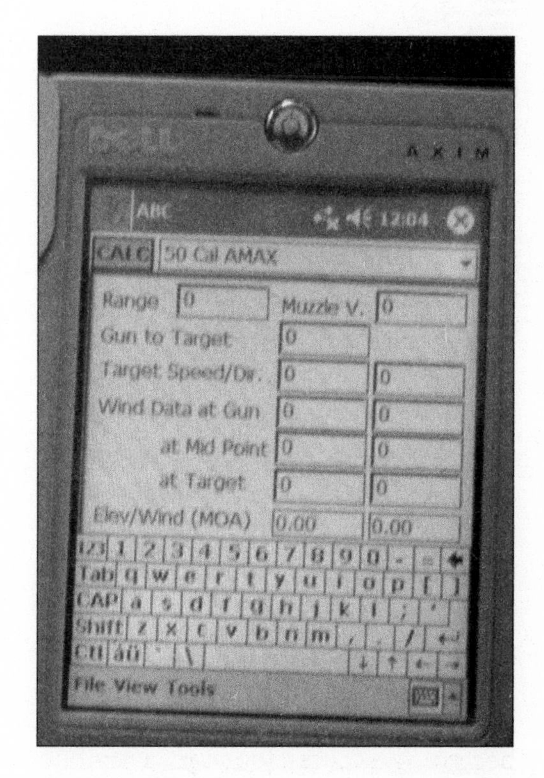


The data will then be input into a device like this PDA, but technology is pushing it into devices that double as tactical watches as well.

3

# Bolt-Action Rifles

The world today is truly a dangerous place. Urban crime, the global war on terror, home-grown terrorists plotting to kill us, the wars in Iraq and Afghanistan, and desperate men holding innocent people hostage—all are evidence of just how dangerous our world has become. In all of these violent situations, there is a leader, someone who is directing forces or holding hostages and demanding terms. In any endeavor, be it a business, a military unit, a pirate crew holding a ship captain hostage, or a crazed man holding his family hostage, if the leader is eliminated, the rest of the organization quickly falters. All of these situations call for the highly refined and unique skills of the 21st-century sniper.

The capabilities of the modern-day sniper are vast. If you think about it for a moment, every job really requires the same basic components: someone who is motivated to do the work, possesses the necessary skills and competence to perform the task, and is provided with the proper equipment to complete the job. If these three components are lacking, the job will either be inadequately performed or not be accomplished at all. If the proper tools are not available, the job

**Coauthor Glen Doherty poses on the set of a National Geographic sniper special in Southern California**

can in some instances be done—but in most instances very inefficiently. You could use a screwdriver to nail a railroad spike, you could use a steak knife to cut down an oak tree, or a Swiss Army knife to attack a charging grizzly, but the outcome may not be what you were hoping for. Understanding this, it is easy to see that today's 21st-century sniper uses specialized technology in addition to his rifle as tools to accomplish his varied missions. If the sniper's mission calls for eliminating a gang member holding a woman hostage in a bank, the use of a .50-cal would get the job done, but would be overkill, especially since the round will continue through multiple unknowns, including the back wall of the bank. Similarly, if the mission is taking out an enemy combatant at 1,500 meters, a .308 would not be the ideal sniper "tool" of choice.

Bolt-action rifles have been in use since the early 19th century, and while they are being slowly phased out by the increased accuracy, practicality, and reliability of the semiautomatics, they remain in heavy use today. The sniper bolt-action rifle comes in several different calibers and designs. Today's sniper has, in many instances, the choice of the "right tool" for the mission, but there are some rifle/calibers that may have broader application than others.

## A Brief History

The bolt action came into existence with the development of smokeless powder in the mid- to late-1800s. Prior to the development of smokeless powder, rifle ammunition was powered by black powder. As most shooters know, ignited, black powder causes a large volume of smoke—not a good thing if you are a sniper, trying to fire upon the enemy without being detected. Johann Nikolaus von Dreyse designed the first crude bolt gun in 1848. In 1866, the French introduced their version

**Friends of the author from left to right: author Billy Tosheff, MLB player Kevin Kouzmanoff, and a SEAL sniper.**

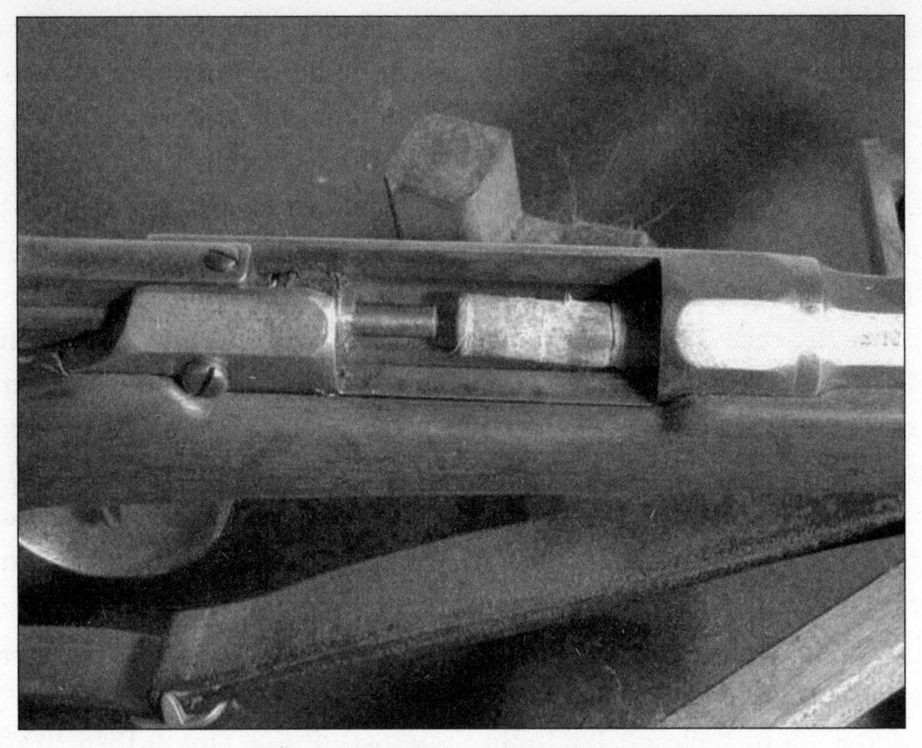

Chassepot "bolt" and chamber with old round

of a bolt rifle, the Chassepot "Needle Gun," officially known as the Fusil Modèle 1866.

These new rifles gave the sniper the ability to fire a round by releasing the firing pin, which allowed the trigger to be lighter, thereby giving the sniper better trigger control. In 1871, the German brothers Paul and Wilhelm Mauser introduced the first true bolt-action rifle, which was the first to use a true center-fire cartridge. At that time, the U.S. Army was using the .45-70 Government (.45-cal bullet

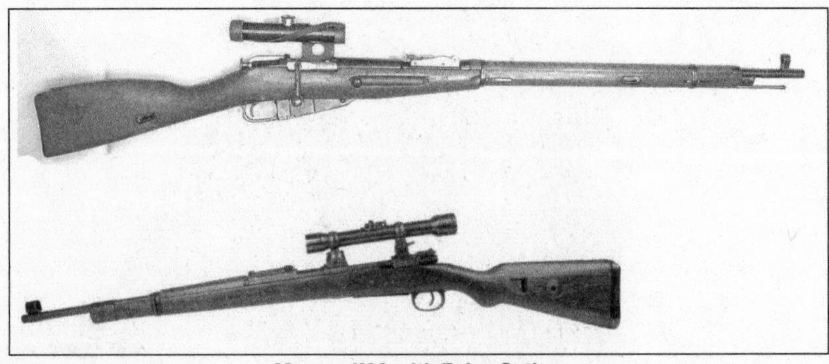

Mauser K98 with Zeiss Optics

with 70 grains of black powder) that generated about 1,300 fps at the muzzle. The French-designed 8x50mm Lebel using a 198-grain bullet generated 2,380 fps at the muzzle, really the original smokeless round to be widely used by any one country. Across the English Channel, in 1888, the British, introduced the Lee-Enfield 303, which was capable of using metallic cartridges and could be loaded with more than one round in the chamber—the first repeater, in that it had a ten-round magazine. In that same year, the Mauser brothers introduced their next generation bolt rifle, the 8mm Mauser, to the German army. The action on this rifle is still used today. The Russians developed the Mosin-Nagant Model 1891, which fired a new rimmed cartridge, the 7.62 x 54 mm.

Like the German Mauser and the British Enfield, it saw service through World War II as Russia's primary sniper rifle. By the late 1880s, the United States recognized that its single shot .45-70 Trapdoor Springfield was not a suitable rifle given the developments of the Europeans. In

**Zeiss reticle**

1890–91, the U.S. Army looked at many rifles and decided to go with a rifle designed by Norwegian army captain Ole Krag and Erik Jørgensen—the .30-40 Krag, which was designated as the Model 1892. The rifle shot a .30-caliber 220-grain bullet with 40 grains of smokeless powder. This load combination generated about two thousand feet per second (fps) at the muzzle, which was inferior to the German, French, and British cartridges but was better than the .45-70. The effective range of this rifle was about 400 to 500 yards, but the best group size was about 3.5 to 4 inches at 100 yards. A horrible 3.5 to 4 minute of angle (MOA)!

In World War I, the German Mausers were the best sniper specific rifles in existence. The Germans were also the first to put optics on their sniper rifles. In the United States, the .30-06 Springfield became the standard U.S. military cartridge. The .30-06 was originally adopted in 1903 along with the Model 1903 Springfield. This cartridge was designated as the .30-03. However, severe erosion of the barrel's throat with the powder being used at the time forced the round load to be reduced, which resulted in a muzzle velocity of 2,000 fps. This was the same as the .30-40 Krag that the original .30-06 was supposed to replace. U.S. Ordnance modified the cartridge case slightly with a 150-grain bullet, which then generated a muzzle velocity of 2,700 fps. The cartridge was then designated the .30-06, which is still the case in use today. The .30-06 saw service throughout both World Wars.

The 308 development began at the end of World War I, but it wasn't until 1957 that the government finally used the 7.62 NATO in more than

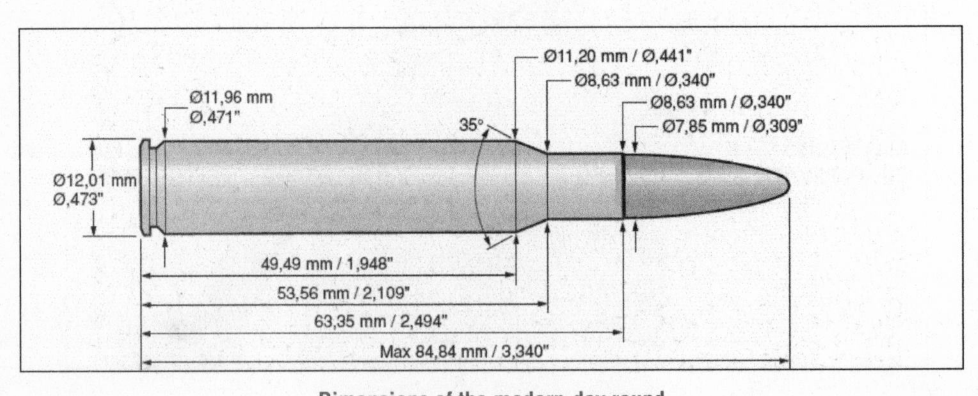


Dimensions of the modern-day round

Photo Credit: CheyTac

The modern face of the bolt action rifle: The Cheytac M200 Intervention .408

one weapon, getting it into the M14 and M60 machine gun.

As developments in metallurgy, powders, bullets, and the science of projectile ballistics continue to progress, other cartridges have been developed. As noted earlier, these new cartridges were developed to provide the sniper with a better tool to accomplish his mission. The 308 remains a very popular cartridge for the 21st-century sniper, in both bolt and semiautomatic versions. It is probably the most inherently

Photo Credit:Smith and Wesson

Smith and Wesson M&P 5.56 semi-auto

accurate .30-caliber commercial cartridge ever produced.

## The Bolt-Action Rifle—The Components and How It Goes "Bang"

Rifles—all guns for that matter—can be classified by the type of "action" that places the cartridge into the firing chamber. Modern rifles have three different actions, a) the lever action, b) a semiautomatic action, and c) a bolt action. In the lever-action rifle, a lever pulls and pushes the cartridge into and out of the chamber. Lever-action rifles are often in old Western movies and are often referred to as cowboy rifles. A semiautomatic rifle, like the M16 or M14, uses the gas generated by the burning powder and springs to move the loading/unloading mechanism. In both of these examples the loading/unloading mechanism may be referred to as a bolt in the owner's manual and indeed it performs the same function as the bolt in a bolt-action rifle.

In a bolt-action rifle, the bolt is moved in and out of the receiver by the shooter's hand. The bolt is pushed and twisted forward to load the cartridge into the firing chamber. To unload the spent cartridge, the bolt is twisted and pulled back to eject the spent round. The twisting of the bolt is similar to the motion or action used to tighten and untighten a bolt, hence the name "bolt-action." So now that we have an idea of what a bolt-action rifle is, lets get to the components of a bolt-action rifle.

All rifles are composed of the following basic components:

A. The action. The action is composed of two parts—the receiver and the bolt. The receiver of the rifle is where the cartridge enters the action. The bolt pushes and pulls the cartridge into and out of the firing chamber. The

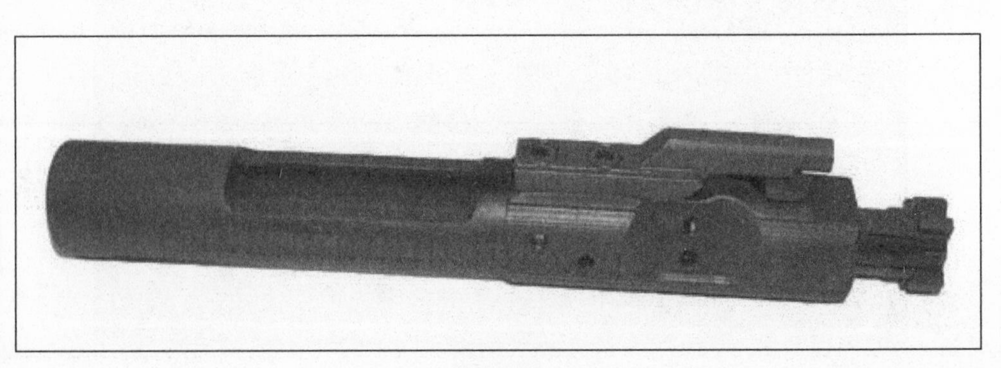

AR 15 bolt

Photo Credit: Benelli U.S.A., Corp.

Details of 1860 Henry Rifle

bolt houses two components—the firing pin and the cartridge extractor. The firing pin is what strikes the primer of the cartridge when the trigger is pulled. The extractor is the part of the bolt that pulls the spent round out of the chamber and ejects it out of the receiver in preparation for the next cartridge. The action plays an integral part in the firing of the rifle. In sniper-grade rifles, the action parts are machined to very close tolerances and should be trued, as even minor misalignments can cause problems with the accuracy of the rifle.

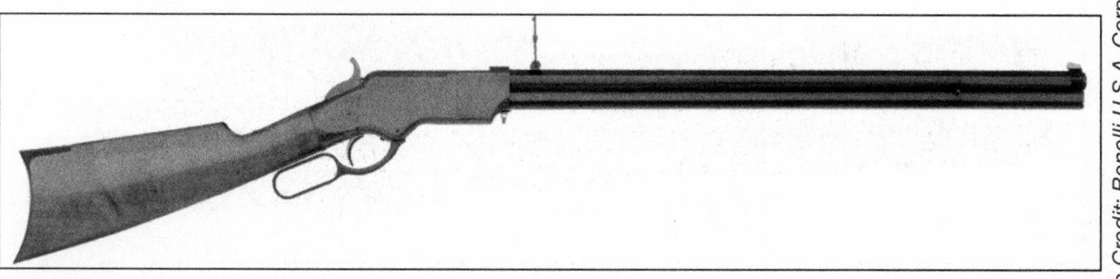

Photo Credit: Benelli U.S.A, Corp.

The Marlin Lever Action rifle

**HS Precision 308**

*Photo Credit: Curtis Prejean, courtesy of Accuracy International*

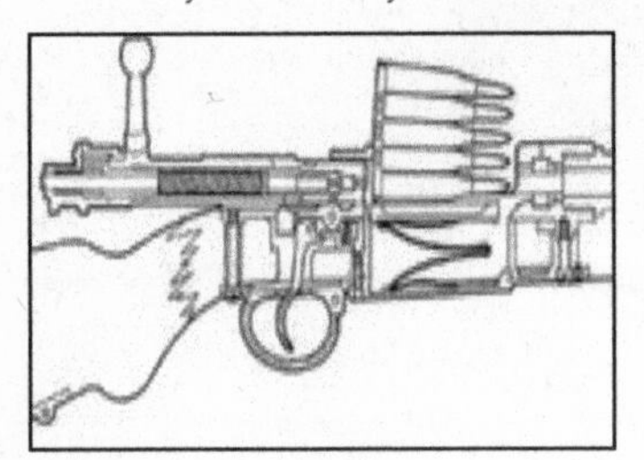

**Cut-through view of the Accuracy International.244**

*Photo Credit: Curtis Prejean, courtesy of McMillan*

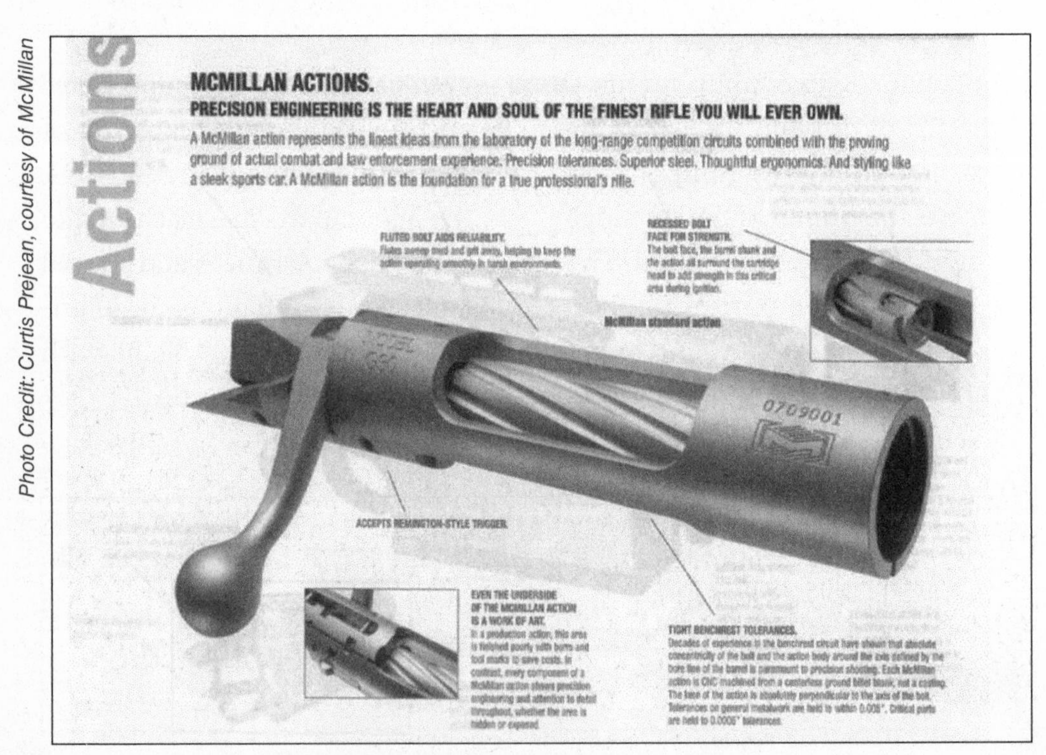


**McMillan action explained**

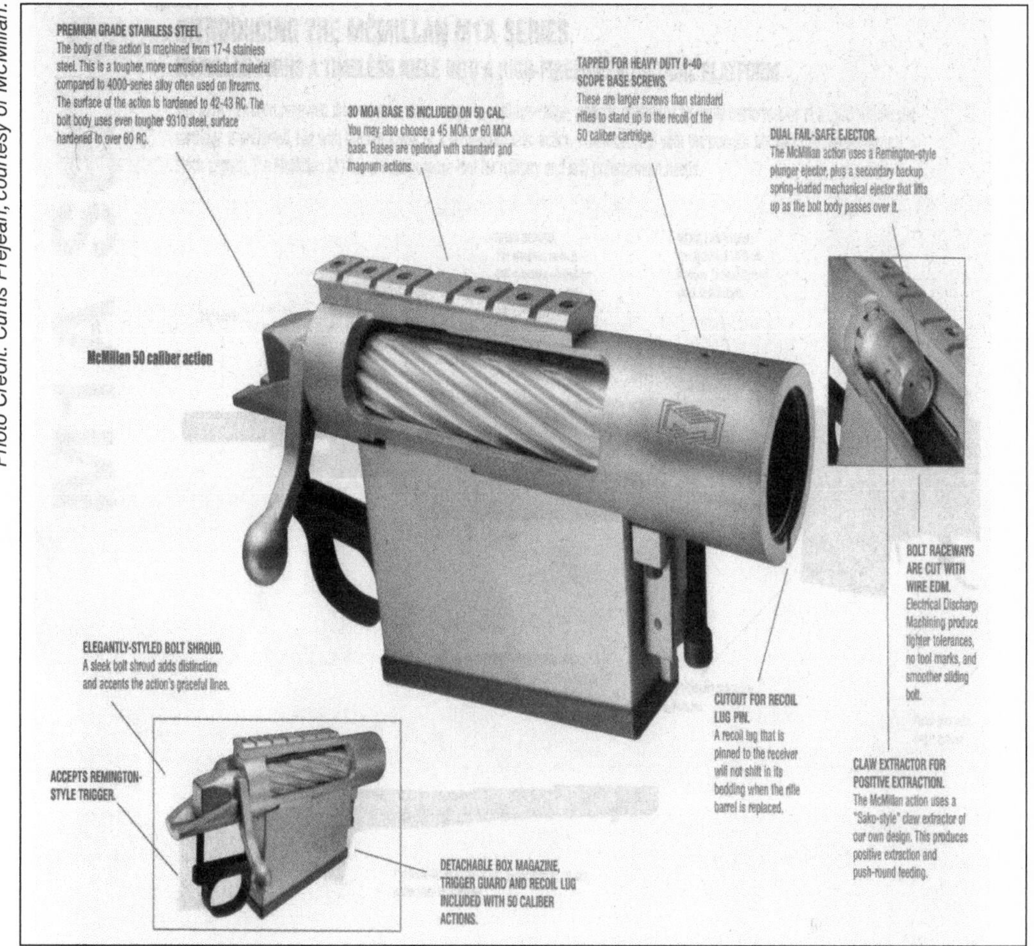
PREMIUM GRADE STAINLESS STEEL.
The body of the action is machined from 17-4 stainless steel. This is a tougher, more corrosion resistant material compared to 4000-series alloy often used on firearms. The surface of the action is hardened to 42-43 RC. The bolt body uses even tougher 9310 steel, surface hardened to over 60 RC.
30 MOA BASE IS INCLUDED ON 50 CAL.
You may also choose a 45 MOA or 60 MOA base. Bases are optional with standard and magnum actions.
TAPPED FOR HEAVY DUTY 8-40 SCOPE BASE SCREWS.
These are larger screws than standard rifles to stand up to the recoil of the 50 caliber cartridge.
DUAL FAIL-SAFE EJECTOR.
The McMillan action uses a Remington-style plunger ejector, plus a secondary backup spring-loaded mechanical ejector that lifts up as the bolt body passes over it.
McMillan 50 caliber action
BOLT RACEWAYS ARE CUT WITH WIRE EDM.
Electrical Discharg
Machining produce
tighter tolerances, no tool marks, and smoother sliding bolt.
ELEGANTLY-STYLED BOLT SHROUD.
A sleek bolt shroud adds distinction and accents the action's graceful lines.
CUTOUT FOR RECOIL LUG PIN.
A recoil lug that is pinned to the receiver will not shift in its bedding when the rifle barrel is replaced.
ACCEPTS REMINGTON-STYLE TRIGGER.
CLAW EXTRACTOR FOR POSITIVE EXTRACTION.
The McMillan action uses a "Sako-style" claw extractor of our own design. This produces positive extraction and push-round feeding.
DETACHABLE BOX MAGAZINE, TRIGGER GUARD AND RECOIL LUG INCLUDED WITH 50 CALIBER ACTIONS.

Photo Credit: Curtis Prejean, courtesy of McMillan.

**McMillan .50 caliber action**

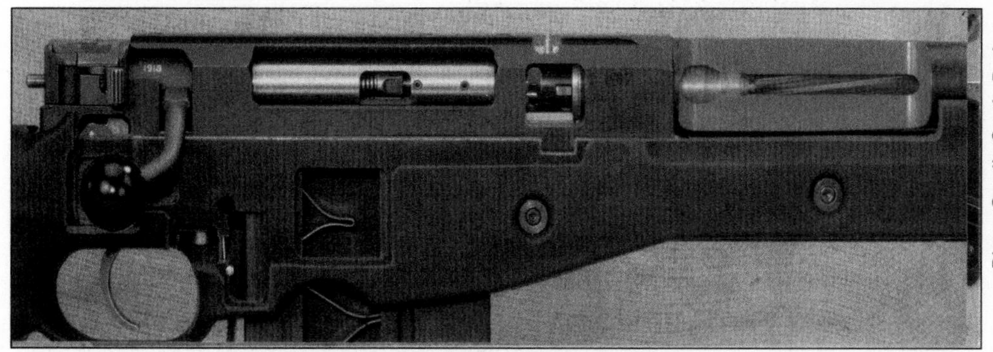

Photo Credit: Curtis Prejean, courtesy of McMillan.

**Cutaway of an Accuracy International bolt action**

Photo Credit: RCArms.com

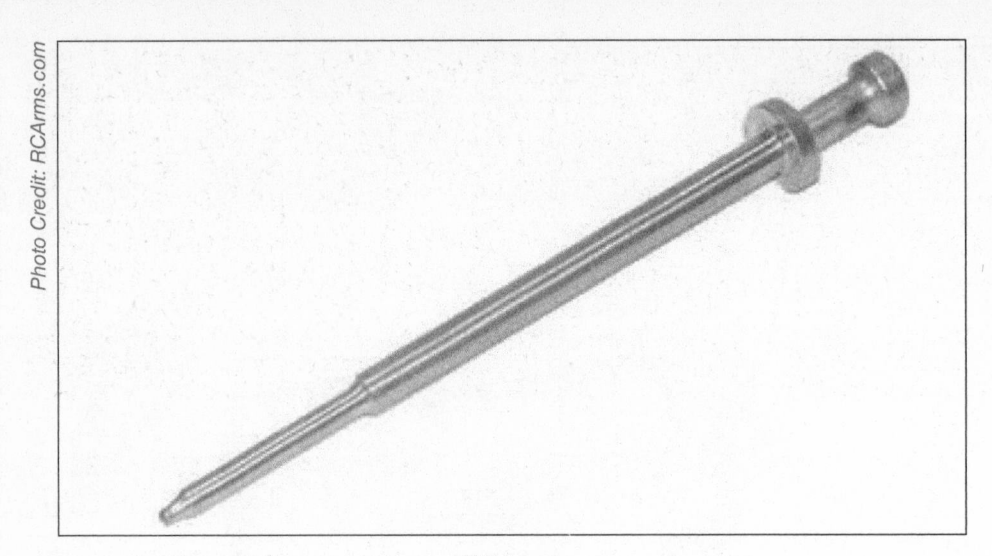

M4 firing pin

The rifling of a barrel

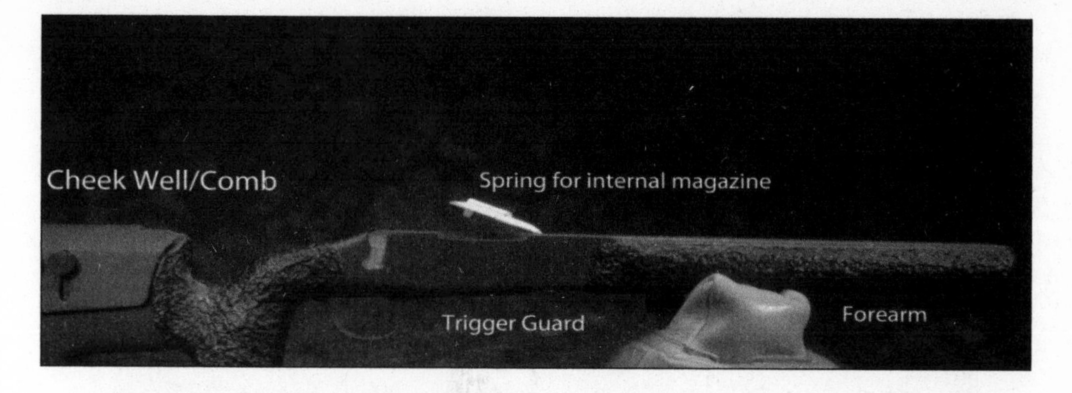


B. The trigger/safety mechanism. The trigger mechanism is made up of several parts that work together to deliver a strike or hammer to the firing pin located (usually) in the bolt. The hammer drives the firing pin forward to the primer of the cartridge. When the firing pin strikes the primer, the explosive component in the primer is ignited, which then in turn ignites the powder in the cartridge case. As the powder burns, it very quickly creates enormous amounts of gas. As the gas rapidly expands, the bullet in the case is forced down the barrel and out to its intended target. There are two types of triggers: two stage and single stage. We will address the differences later in this book. In sniper-grade rifles the trigger must be very smooth, dependable, safe, and easy to manipulate. A smooth trigger mechanism is a key component in building an accurate sniper rifle.

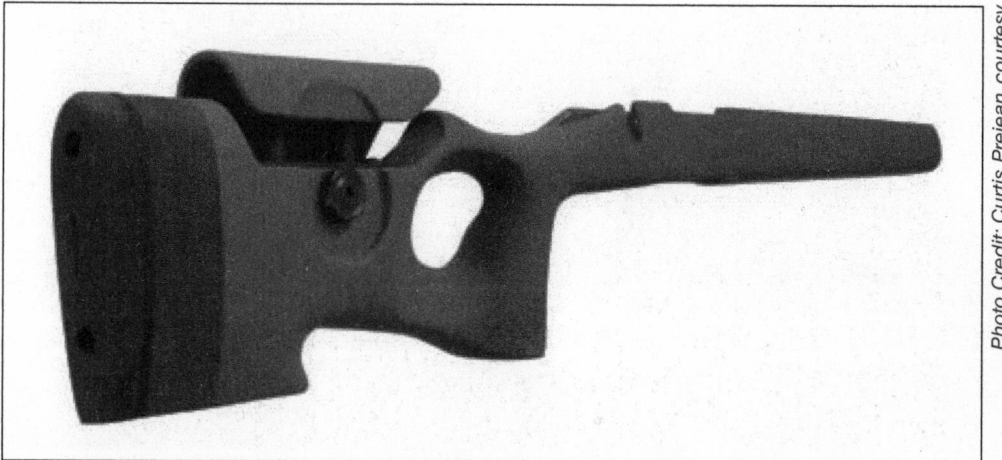

*Photo Credit: Curtis Prejean, courtesy of McMillan.*

**McMillan stock**

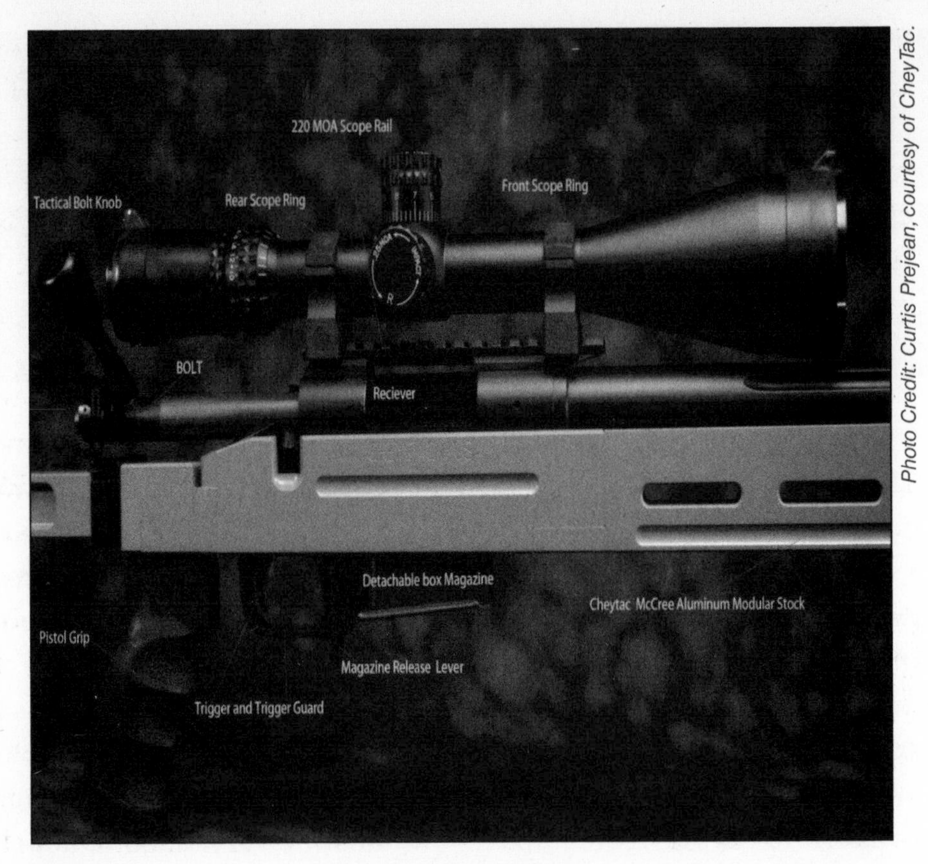


Photo Credit: Curtis Prejean, courtesy of CheyTac.

**CheyTac defined in detail**

**Modern-day bolt-action. Notice the adjustable cheek weld and rail system for mounting additional lasers and optics.**

C. The barrel. The barrel is probably the most important factor influencing the inherent accuracy of a rifle. Barrels and their characteristics will be discussed in greater depth later in the chapter. The barrels are where the bolt pushes the cartridge into the part of the barrel known as the chamber. The chamber then narrows to the designated caliber diameter. All rifle barrels have spiral grooves cut into them that engage the bullet as it travels from the cartridge to the end of the barrel. These grooves are known as rifling. The spiral curve of the rifling causes the bullet to spin, which helps give it gyroscopic stability, which greatly increases the accuracy of the projectile. The science and art of building an accurate barrel is complicated, as many factors influence the barrels accuracy.

D. The stock. The stock is the part of the rifle that enables the sniper to hold on to the rifle. It can be made of wood, fiberglass, aluminum, or a combination of materials. The modern sniper rifle is usually made of fiberglass/synthetic material or aluminum.

These four components make up all rifles. There are other accessories that must be added to the rifle to make it functional. The first required accessory is some kind of sighting device. Sight options can vary from the basic iron sights to optical telescopic sights. Modern sniper rifles are all equipped with optical telescopic sights so discussing iron sights with sniper bolt-action rifles is pointless—they don't have them. Nonrequired sniper rifle accessories include a muzzle brake, a bipod, and perhaps a suppressor. Suppressors are discussed in another chapter, so we won't delve into that topic here. The same is true for telescopic sights—you can jump to that chapter when you are done here.

4

# The Way of the Future: Semiautomatics

**Semiautomatic**: *a firearm*: able to fire repeatedly but requiring release and another pressure of the trigger for each successive shot.

Let's first take a look at the history behind the semiautomatic rifle. There were a few early attempts, but a German gunsmith named Ferdinand von Mannlicher invented the Model 85 in 1885, the first known successful semiautomatic rifle. American John Browning followed up a few years later with the first successful semiautomatic shotgun (Browning Auto-5). The Auto-5 relied on long recoil operation; this incorporated coupled the recoil action with a spring to re-chamber the round and was a design that was in use for over fifty years until production ceased in 1999. Winchester was the first to introduce the semiautomatic rim-fire and center-fire rifles into the civilian market in 1903 and 1905.

Surprisingly, the first country to incorporate a semiautomatic rifle was France! The rifle was the Fusil Automatique Modèle 1917. This was a recoil-operated action rifle used in World War I. Toward the latter part of World War I, several countries including the Soviet Union, Germany, and Britain had started to incorporate the semiautomatic

The new badass Springfield Armory SOCOM II

into their inventories. Britain's goal was to replace the very reliable bolt-action Lee-Enfield rifle. I have personally run across a few of the Enfields still in use in Afghanistan, a tribute to the solid construction and reliability!

The first semiautomatic that became standard issue for the U.S. military was the M1 Garand, a gas-operated rifle developed in Canada by John Garand for Springfield Armory. The first production model was released in 1937. The semiautomatic proved itself under fire in World War II and gave the U.S. soldier quite an advantage over German opponents who were still mostly using bolt-action rifles. It seems a bit ironic that the country that invented the semiautomatic didn't incorporate them into their military in larger numbers earlier!

## Birth of a Legend

The Soviets are known for their rugged and virtually bulletproof engineering standards, whether

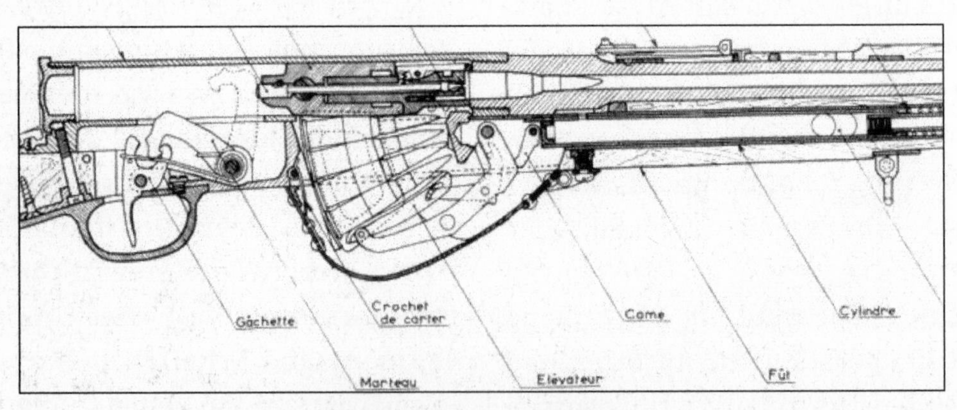


Fusil Automatique patent drawing

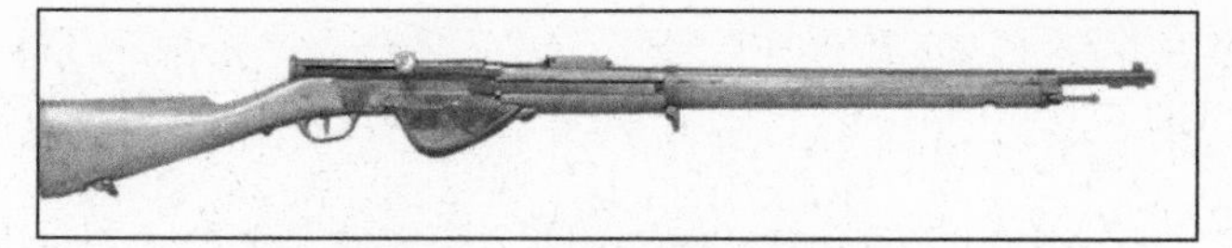

The 1917: a weapon ahead of its time

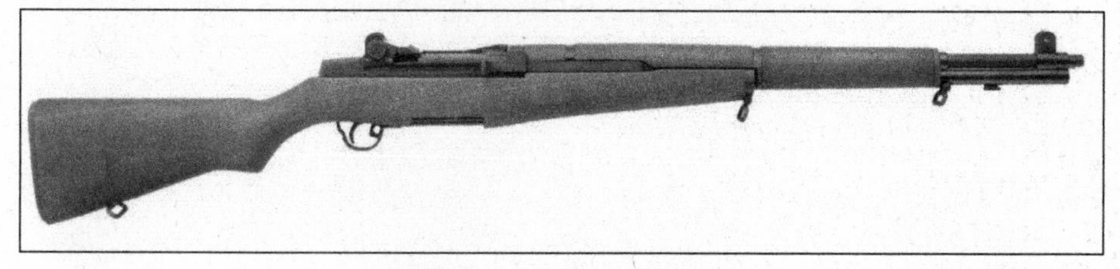

M1 Garand: a battle-proven rifle

it's airplanes, tractors, cars, or guns, and the AK-47 is arguably the most popular and widely-used semiautomatic rifle today. The AK-47 will stand up to the harshest environments, whether it's the frigid mountains along the Afghanistan/Pakistan border or the hot, dusty streets of southern Iraq; this weapon will fire without malfunction. I can personally attest to having used this weapon in training over an eight-hour period, swimming it back and forth through heavy ocean surf and having the weapon fire flawlessly. This rugged and simple engineering is why this weapon is one of the most popular in the world.

Captured selection of semiautos in the early days following 9/11 in Afghanistan

The AK, weapon of choice, is shown here with the transportation of choice in more countries than any other as friendlies get ready to roll in Afghanistan.

## Paradigm Shift

Bolt-action rifles are still employed in modern warfare, but I am here to tell you that the bolt-action rifle is a thing of the past—like a manual transmission compared to the modern-day automatic. While some people will prefer the manual over the automatic, it is clear that the automatic is the future and here to stay and that the manual is fading quickly. The same can be said of the semiautomatic rifles of today: They are just as reliable and accurate (subminute of angle accuracy) as a good bolt action and a hell of a lot more practical to the modern-day sniper. That said, there has to be some discussion of quality: Not all semiautomatics are created equally and there is plenty of garbage out there. For a while in the SEAL teams, we were using the Knight's Armament 7.62 SR25. To be honest, I am not a big fan, as I've experienced critical malfunctions personally and seen more than a few of these rifles have total mechanical failure in training. When you are in an urban hide in Basra,

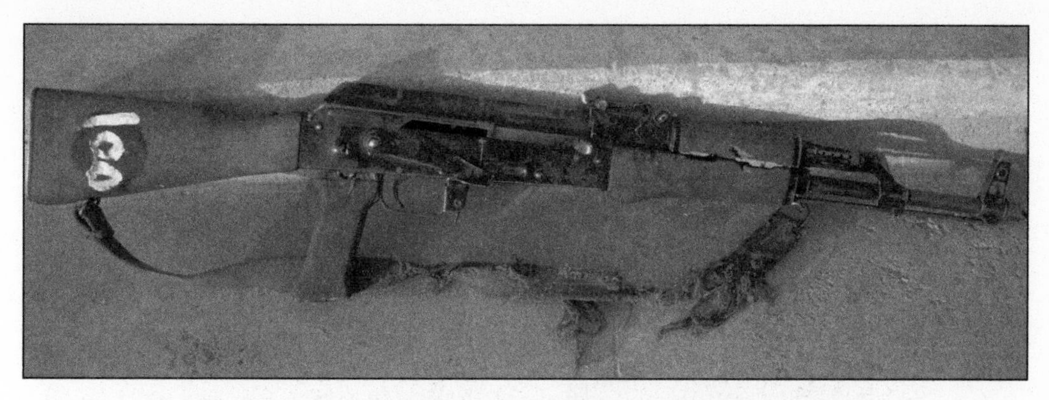

"Only dropped once." Note the blood on the foregrip and the two bullet holes through the weapon, one on each grip. Taken off a fallen Iraqi soldier outside Tikrit, Iraq.

Iraq, surrounded by bad guys that want nothing more than to string you up by the neck, having a malfunction is unacceptable.

It is very unfortunate that there are very few American weapons manufacturers who are currently making a full-scale production model semiautomatic that can compete in the global market. (Note: The authors love the new Springfield SOCOM II.) Like the collapse of the American auto industry, the harsh reality of it all is that there are few, if any, U.S. weapons manufacturers out there that are making high quality products and pushing the technology envelope. The Europeans have been kicking our ass with their advanced engineering and forward thinking. I will say that Smith & Wesson Holding Company is beginning to catch on, and in my conversations with management, they say they are positioning themselves for change into the new century. This is refreshing to see from a U.S. company.

Snipers in today's modern-day warfare value the semiautomatic for its reliability and accuracy, but also as a weapon that can be used to engage the enemy in a sustained

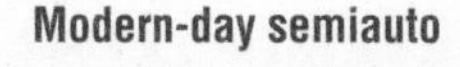

Modern-day semiauto

Brandon Webb training with Kurdish commander in Northern Iraq.

HK 417 with a U.S. Optics SN series variable power scope. A sweet shooting 7.62 that can reach out and engage multiple targets quickly the way only a semiauto can. Note the match grade barrel shown next to a box of LeMas Ltd's BMT 7.62 match ammo.

The Remington Modular Sniper Rifle

firefight and serve as a battle weapon in close quarters when clearing multiple rooms in a building. There is no substitute for the versatility and adaptability of the modern-day semiautomatic. In Iraq our SEAL sniper teams were devastatingly effective with the semiautomatic. We owned the urban environment and literally shut down miles of city because of our expertise coupled with the rapid accurate fire that our semiautomatics were able to deliver. We have some of the highest kills in the Special Operations community because of our high training standards coupled with the modern-day semiautomatic.

## Great Modern-day Examples

Modern semiautomatics fall into three categories of operating systems, with several subcategories for each:

1. Blowback operated. In most designs, the bolt is actually unlocked at the moment of firing. There is a delay mechanism used to prevent excessive force from operating the bolt; your kid's Ruger .10-22 is an excellent example of this action. For the grownups, the H&K PSG-1 uses a roller delayed blowback design to completely close the bolt face prior to firing, increasing the accuracy to

**Coauthor checks out the feel of the suppressed HK 417.**

Photo Credit: HK

Heckler and Koch PSG-1 Counter Sniper Rifle

sub-minute of angle and increasing the range to 800 meters.

The roller delay mechanism has been around for decades and was first deployed by Mauser in 1945. The mechanism forces the bolt carrier rearward at high velocity and leaves the bolt face in place, letting the mass of the bolt provide inertia that creates a time delay required to ensure that the bullet has left the barrel and the gas pressure in the barrel has dropped to a safe level. It's still unpleasant to stand next to. If the hot casing doesn't hit you, the gas from the chamber will. Many sniper quality semiautomatics fall into this category.

2. Recoil operated. The bolt is not completely engaged in recoil-operated weapons and is held closed by a spring. When the weapon is fired, the recoil overcomes the spring tension and the bolt may cycle. The spring tension will decrease with use over time and the force on the cartridge will decrease the force on the bolt face will become inconsistent and the effect on the bullet trajectory will be drastic.

Recoil operation is usually limited to smaller caliber weapons due to the amount of force transferred to the weapon frame and the operator by the moving bolt.

3. Gas operated. Gas operation of self-loading firearms has been around for more than a century. The early designs were often prone to failure. The first design that was considered useable was John Browning's in 1891, with the first working prototype able to fire sixteen .44-caliber rounds in under one second!

There are two types of gas-operated mechanisms. The first uses a very stiff spring to keep the bolt face closed and a small amount of gas is tapped from the barrel to push the bolt back

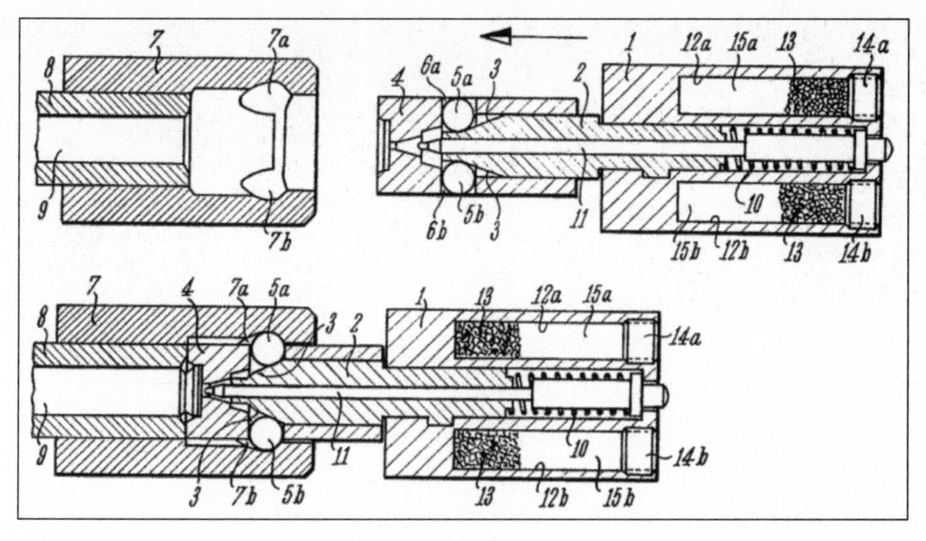


**HK is famous for the reliability and the simplicity of their roller delayed operation shown here in detail.**

against the spring; the AK-47 uses this method, as it has the advantage of being very inexpensive and simple to manufacture.

The second type is the direct impingement method, where the tapped gas is vented into a tube on the bolt directly and blows the bolt backward after firing, cycling the weapon. The M-16 uses this method of operation with the addition of a

*Photo Credit: Department of Defense Visual Information Center*

**Traditional shooter/spotter pair using an M21 on the left and an M24 on the right. Most 21st-century snipers are operating in much closer environments as lone gunmen. This is a departure from the past.**

rotating bolt to increase accuracy.

There are a large number of variations, such as the tilting breechblock used by Fabrique Nationale de Herstal (FN) in their FAL weapons to create a consistent bolt face closure and increase accuracy and range. The FN-Special Operations Combat Assault Rifle (SCAR) uses a gas-operated rotating bolt to achieve a respectable 625 rounds per minute cycle rate, with subminute of angle accuracy and 800-meter range for the 7.62X51 NATO cartridges.

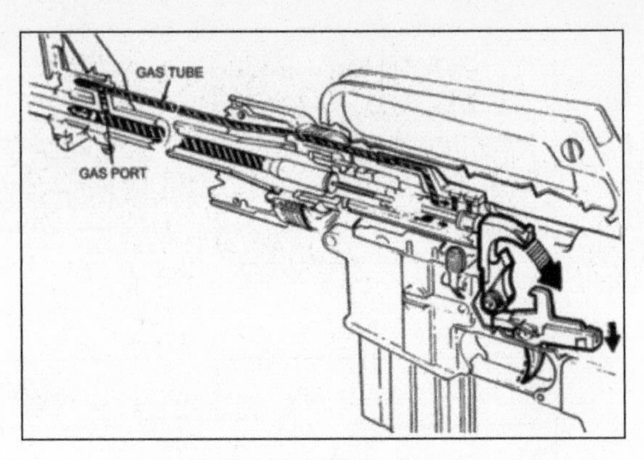

Direct Impingement Gas Pathways can lead to a lot of fouling and carbon buildup in the chamber.

The two great examples of modern-day subminute of angle semiautomatics that we will look at in this chapter are Heckler and Koch's MSG90A1 and the FN's Special Operations Combat Assault Rifle (SCAR light and SCAR heavy) system. The FN system was selected by U.S. SOCOM for use by all U.S. Special Operations forces.

Heckler and Koch is a German weapons manufacturing company that is most famous for their MP5 submachine gun commonly referred to as the room broom by Special Operations personnel. The company was the first to use synthetic polymers in their weapons, which reduced weapon weight significantly.

The MSG was Heckler and Koch's answer to a militarized version of the PSG-1 and is a very reliable finally engineered subminute gun. The barrel is weighted precisely at the muzzle to increase the harmonic stabilization of the barrel whip, which increases the accuracy.

I can honestly say that in over a decade in the SEAL teams I have never had a Heckler and Koch weapon malfunction on me.

## FN—Special Operations Assault Rifle

After achieving great success in 2007 in a limited competition between the M4 Carbine, the FN-SCAR, and the previously discontinued H7K XM8, the FN-SCAR-L will be replacing the M4 Carbine and the FN-SCAR-H will be replacing the M14 and MK11

Photo Credit:HK

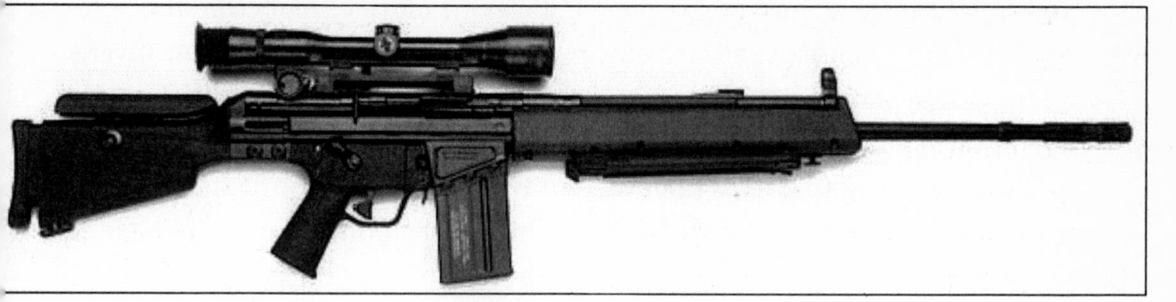

HK MSG90A1. Headache to follow!

sniper rifle. The FN SCAR system offers the user the option to switch caliber with ease and that is invaluable as it simplifies things greatly and gives the sniper choices both urban and rural. The company produces a SCAR light (5.56) and a SCAR heavy (7.62) version.

The main advantage you have in my opinion is a huge increase in the capacity to deliver rounds on target. This capacity to take down targets in very rapid succession is critical when shots are snapping over and around you in the heat of battle. In addition, you have a versatile weapon that can be used in close-quarters battle and helicopter-sniper operations. Couple this with modern reliability and you can say good-bye to the bolt-action rifle.

It is likely that you will have a few hardcore holdouts that will insist this is not the case and that nothing takes the place of the bolt action. But this isn't true. I would challenge you to ask these holdouts if they have recent experience on the asymmetric modern battlefield.

The modern-day semiautomatic sniper has proved its accuracy and reliability on today's battlefields over

FN SCAR with Night Force Optics . . . great combination.

and over again. It has performed reliably and taken thousands of souls on the complex urban streets of Iraq and in the cold mountains of northeastern Afghanistan. The semiautomatic is here to stay and it will only be a few more years before the bolt-action rifle is phased out of the inventory all together. The semiautomatic has earned its place rightfully in the inventory . . . until the next innovative shift in weaponry occurs.

Photo Credit: FN U.S.A.

A CH-53 comes in to extract the author's SEAL platoon from the mountains in Afghanistan.

Not exactly sniper rifles, but definitely an innovative shift in weaponry and a good urban sniper sidekick. The FN F2000 series is a well balanced carbine that can be tailored as the primary weapon for a variety of missions.

"When I was in northern Afghanistan conducting reconnaissance and patrol missions I was torn between the reliability of my 300 Win Mag and the comprehensiveness and rapid-fire capability of my SR-25 semiauto. The problem at the time was that I did not trust my semiauto to hold up to the harsh conditions because I'd had malfunction problems with the Knight's Armament weapon in the past. This has been the reason bolt-action rifles have long been the sniper's choice because of their trust and reliability in the bolt-action system. I ultimately chose to patrol with my M-4 and carry my 300 Win Mag on my back. With the reliability and accuracy of semiautos of today, I would choose a semiauto in this situation, hands down. I would have been lighter on my feet and had a much more efficient weapon system."

# 5

# Everything Else You Really Need to Know About Rifles

The barrel is probably the most important factor in determining a rifle's inherent or potential accuracy. A basic knowledge of what makes a good barrel will help you understand what it takes to build a great shooting rifle. Understanding barrel characteristics will help you realize what your rifle can do and why, as well as diagnose any accuracy issues that arise. Additionally, you will understand why such things as "a free-floated barrel," "don't rest your barrel on anything," and a "fast enough twist" influence a shooter's ability to put bullets on target.

## Barrel Considerations

A nonshooter will look at a rifle barrel and see a round metal bar with a hole drilled down the center. They will probably realize this is where the bullet leaves the rifle. However, the science involved in building an accurate sniper rifle barrel is much more than simply drilling a hole down a metal bar.

*Photo Credit: Biz Times and Krieger Barrel*

The barrel's length, the type of steel or material used, the rifling, the twist rate, and weight all play very important roles in the accuracy of the barrel.

## "It's the steel, my son."

—*Conan the Barbarian*

Today most rifle barrels are constructed of some kind of steel alloy. Alloy steels contain elements that are added to modify the behavior of the steel during the heat-treatment portion of the manufacturing. Nickel increases hardness and tensile strength. Vanadium adds the ability to resist repeated stress. Tungsten adds air-hardening qualities. Molybdenum improves high temperature, service wear toughness, and hardness. Chromium improves the stiffness and hardness. The most common steel in rifle barrel building is known as chromoly. This type is very high in strength if manufactured correctly—tensile strengths will vary from 98,000 pounds per square inch (psi) to 180,000 psi! It also contains chromium, manganese, molybdenum, phosphorus, sulfur, and silicon, but no nickel.

The other type of steel used in rifle barrels is stainless steel. Stainless costs more than chromoly. Stainless was originally used for knives and was patented in 1916, but the name was not trademarked, so now the "stainless" designation is used for any steel that resists rust, corrosion, and acids. There are over 50 different stainless steel compositions, and all have their unique properties. In order to get the stainless steel to have the required tensile strength for rifle barrels, the steel will rust under some conditions. Stainless is good at reducing barrel wear and fouling. One should note that the tensile strength of stainless is lower than properly processed 4140 chromoly steel.

In the manufacturing process the steel goes through a variety of heating and cooling stages to condition the steel to optimize its performance. A

A "gang broach" for rifling barrels. This method of cutting the rifling into barrels is effective, but making the broach and keeping it sharp are difficult machining operations.

**Beautiful rifled shotgun barrel. Most shotguns are smooth bore.**

good steel barrel should be stress relieved during the manufacturing process. Stress relieving is a process of heating the barrel then cooling it in a controlled manner. This process allows the steel molecules to relax and align themselves, which helps reduce any imperfections in the trueness of the barrel.

Once the steel blank is properly processed the barrel must be drilled or reamed. Usually the reaming process takes several passes to bring the barrel to the correct diameter. The outside of the barrel has to be machined to the correct size and should be concentric to the bore of the barrel. The rifling process can

**Rock River Arms Tactical CAR A4 with a 16" Chromoly Barrel**

*Photo Credit: Rock River Arms*

be done in several different ways: hammer forging, button cut rifling, and single cut rifling.

In hammer forging, the process involves the reamed barrel being heated. A hard mandrel is inserted. The barrel is hammered into shape around the mandrel, which leaves the rifle grooves once it is removed. This is usually thought of as the least accurate method of rifling, but it is the cheapest, and several manufacturers employ this method in all of their rifles with good accuracy.

In the process of button cut rifling, a barrel blank is center bored and a hard metal button with the rifling ridges is pulled through the barrel, producing the rifle lands and grooves. Button cut rifling is more expensive than hammer forging and in most rifle building circles is thought to be more accurate.

Single cut, also known as hook rifling, involves the same center boring process, but the rifling is slowly cut with a single groove tool. Obviously, this process is very time consuming since it takes multiple passes for each groove. This process causes minimal stress to the barrel and if done properly is thought by many to result in a very accurate barrel. It is also the most expensive rifling process.

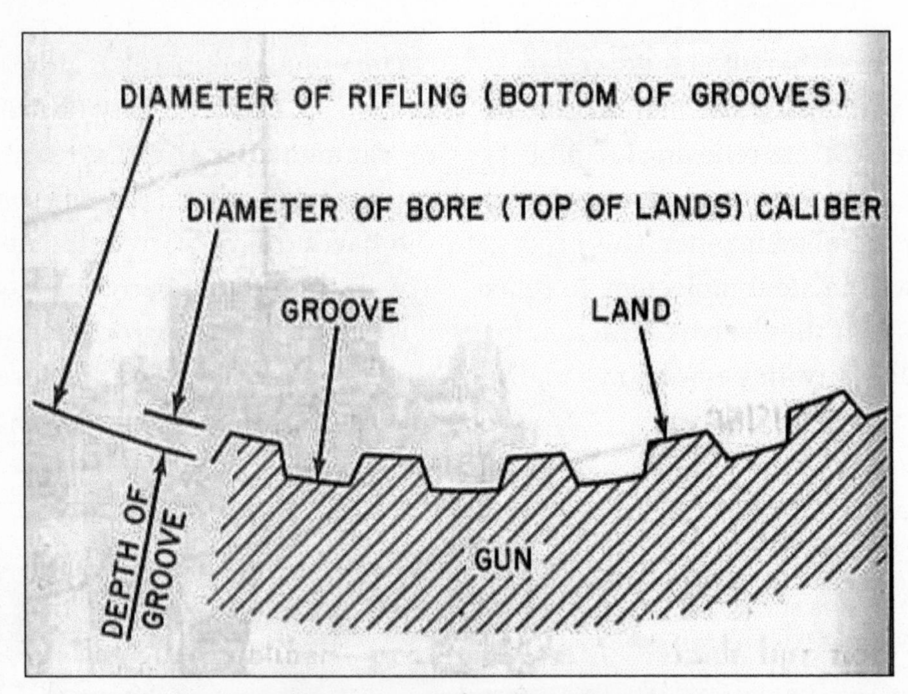


Diagram breaking down rifling grooves and lands. The caliber of the barrel is measured from the internal diameter starting with the top of the lands.

## "Size matters"—Most of the Time

The length of a barrel is also an important consideration in building a sniper rifle. Some factors regarding barrel length are obvious. A long barrel is more cumbersome to handle in confined quarters, heavier to carry on long missions, and may make shooting in anything but a supported or prone position impractical. On the upside, longer barrels are generally more accurate than shorter barrels, especially at longer ranges. The improved accuracy with a longer barrel is secondary to the added weight—less recoil, less movement when the rifle is fired, and higher bullet velocities that result in a flatter trajectory. The higher velocities produced by longer barrels are generally true, which also implies that they are not always true.

When the powder in the cartridge is ignited, the resulting hot gases expand 800–1,300 times. Since the gas is contained in the brass case, there is only one way for the gas to escape—by pushing the bullet down the barrel. The barrel should have nothing in it except the bullet and the push of expanding gas. The burning of the cartridge powder should occur in the cartridge case, not in the barrel. Interestingly, despite the fact that shorter barrels produce less bullet velocity, the drop in speed and accuracy is small enough that it is of little concern to most hunters. Hunters that seldom shoot over 100 to 150 yards will not notice any discernable drop in accuracy or velocity in a 20- or 22-inch barrel compared to a 24–26-inch barrel. The maximum velocity for any barrel length will vary with the powder, bullet, cartridge, and other factors. After leaving the muzzle of the barrel, the bullet immediately begins to lose speed,

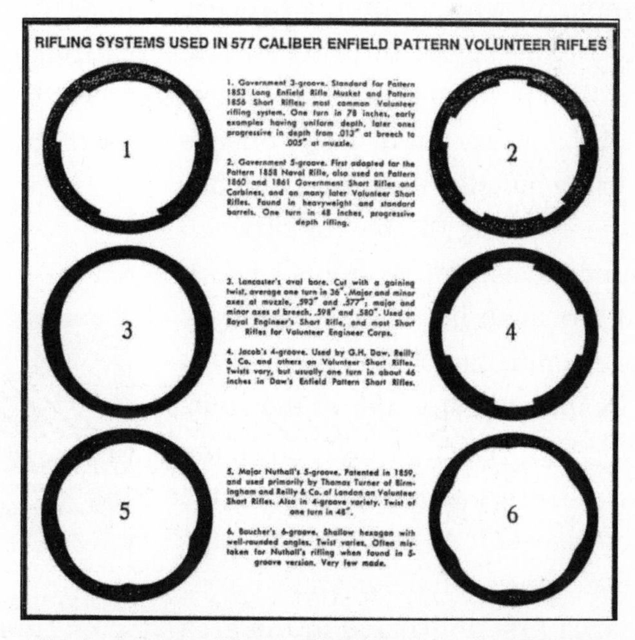


Old chart showing different rifling patterns for the Enfield rifle

**50 cal.**

hence the highest speed of the bullet is directly at the muzzle.

The bore of the barrel should be as smooth as possible. The reaming and rifle cutting process will leave small irregularities that need to be removed by either smoothing the barrel with a lapping process and/or the break-in process. The break-in process varies with barrel manufacturers but usually consists of firing bullets in a certain number while cleaning the barrel between rounds. This process helps smooth out the small imperfections left by the reaming and rifling process.

Despite the strength of the steel, firing high velocity bullets down the barrel causes all barrels to eventually wear out. As the barrel wears, the accuracy will deteriorate. Barrel wear is first noticeable in the area immediately in front of the chamber. This is known as the throat of the barrel. Generally, higher velocity cartridges wear barrels faster than slower cartridges. Depending on the cartridge, the load used to propel the bullet can last from three thousand rounds in an ultrahigh velocity wildcat cartridge to ten thousand rounds in a .308 barrel. Keeping an accurate log of the rounds through a barrel is

*Photo by Curtis Prejean, courtesy of Accuracy International and Tactical Operations*

**Tactical Operations .308 X-Ray in an Accuracy International Aluminum Chassis**

Snipers aren't always on the gun. Here a student sets up his camera for some long-range photographs.

Here a SEAL sniper candidate takes long-range digital photos to be sent back to a Tactical Operations Center for intelligence purposes.

A SEAL sniper candidate veges up his tripod and camera to avoid detection yet still get the intel.

Never underestimate the sniper's importance to reconnaissance and surveillance.

Instructors brief snipers on the "rules" prior to beginning a stalk.

A student tries to locate the OP (observation point) without getting spotted.

Not always warm and dry

Stay low and use the dead space created by the tree to avoid detection.

Find the shooter . . .

Good camouflage

Instructors man the OP table, using high-powered optics to try to spot the students . . . and fail them.

Rain or shine, sniper candidates spend hundreds of hours on the range.

important because of this fact. As a barrel ages, a shooter will notice the group spread progressively getting wider.

## "Twist"

Rifling is the grooves cut into the bore of the rifle barrel. The rifling is cut in a helical or spiral pattern over nearly the entire length of the barrel after the chamber.

Most barrels have a small section of the barrel after the chamber that is known as the free bore or jump. This portion of the rifle does not have rifling. This jump length is very short and in some custom rifles can be eliminated, but more about this later.

Muskets of old and shotguns do not have rifling. They are referred to as smooth bore. The addition of rifling to barrels was first found in the late 1400s and early 1500s. The rifling grooves and their helical (spiral) shape cause the bullet to turn on its axis as it travels down the barrel and then spin at a high rate of speed as it exits the barrel on the way to its target. This spinning motion greatly enhances the bullet's stability just as a perfect spiral motion enhances the accuracy and flight of a well-thrown football.

The amount of twist is important to the accuracy of any bullet. Small differences in twist rate can have major accuracy influence. A good rifle barrel, with a good quality bullet, can perform magical things with the right twist. There are many theories on what twist is best for different-sized bullets. Today, most people acknowledge that there is an ideal twist for each caliber, bullet, and barrel length, but when in doubt, it's better to go for the higher twist rate. For hunting and all around rifles, the twist should be chosen that will work satisfactorily with the biggest and heaviest bullets

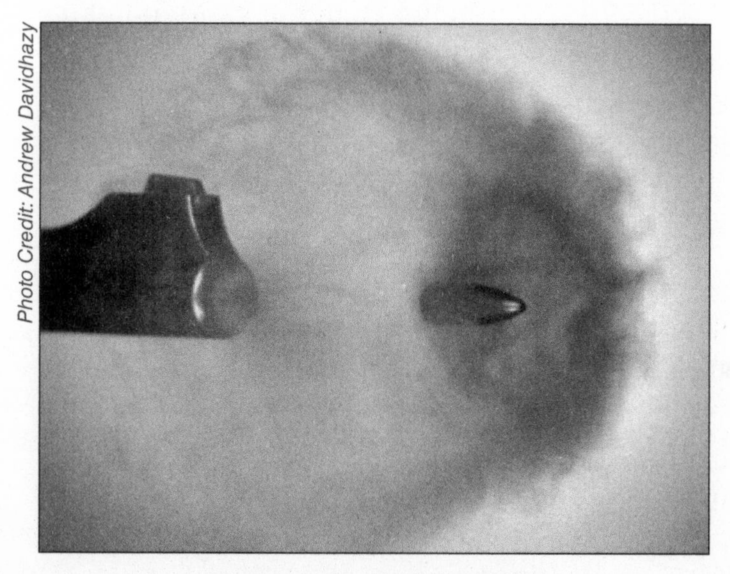

Photo Credit: Andrew Davidhazy

Internal to external bullet transition

made for that rifle. For precision work, the twist should be chosen that will best match and meet the needs of the bullet chosen to do the work. Fortunately for all of us there are smart engineers and gunsmiths working this out for us, so we don't have to lose sleep at night wondering if our twist ratio is appropriate for our ammo.

Proper twist rate requires that we consider multiple factors. In general, to achieve the correct twist we must consider the following: a) a bullet that is longer in proportion to its diameter will require a faster twist to stabilize it; b) long-nosed bullets, such as hollow point boat tails, are good to reduce drag and have increased aerodynamics, but they are harder to stabilize; c) the density of a bullet is also a factor; longer and heavier bullets can be stabilized with a twist suitable for a lighter bullet if we can increase the velocity of the bullet; and d) a slow bullet requires a faster twist.

Other "twisted" facts about twist:

1. If the rifling lands extend to the end of the chamber, less gas escapes and pressure is increased. This permits a lower powder load for the same velocity.
2. Long, pointed bullets are harder to stabilize because any error in manufacturing will cause problems with the center of gravity of the bullet and affect its gyroscopic stability in flight.
3. Plain, short bullets work best with a slower twist.
4. If two bullets are fired simultaneously, with the same velocity and spin rate, the heavier long bullet will keep its speed of rotation better than the shorter, lighter bullet.
5. The best rifle twist is governed by the bullet's length, not its weight.
6. Barrel manufacturers have tolerances for the actual twist, i.e., a 1 in 12 twist may be a 1 in 11 or a 1 in 13. Custom barrels can be made more precisely, but at an increased cost.
7. Increasing twist for a faster rotation speed will not give a flatter trajectory.

8. Rechambered rifles commonly do not shoot as expected because the twist is not correct for the new caliber.
9. If the twist was not correct for a bullet, switching to a heavier, longer bullet will not improve accuracy because the twist is too slow.

## From Trigger Pull to Bullet Flight

As noted above, a tremendous amount of science goes into building an accurate sniper rifle. To be a top sniper does not require that one know or understand all the nuances of the sciences of metallurgy, propellant (powder) chemistry, or ballistics. However, the more shooters understand the science, the better they will understand why certain shooting facts make sense and in turn use these facts to become better. There is a tremendous amount of science involved in building a great bolt-action sniper rifle. The same can be said for what happens to the sniper's rifle once the trigger is squeezed.

The phenomena described below all happen in the microseconds following the firing pin striking the primer, so it's nearly impossible to really "see" these events.

Once the firing pin strikes the primer, the primer compound explodes and ignites the powder in the cartridge case. The resulting hot gas expands in all directions—not just against the bullet's base and the

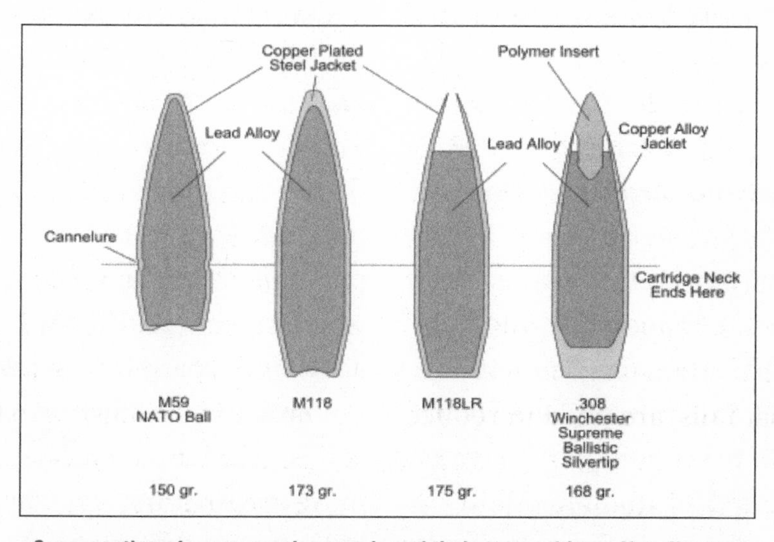


**Cross-section of some popular rounds and their compositions. Note the two rounds on the left are full metal jacket, whereas the other two are considered 'open tip.' There really isn't a lot of difference between the two M118 rounds, aerodynamically speaking.**

Photo Credit: Matt Johnson

**Probably not too concerned about rifle twist this split second. SEAL sniper Matt Johnson gets some trigger time in.**

rear of the rifle chamber. The pressure causes a barrel expansion. The amount of expansion or swelling of the barrel depends on the type of barrel, its thickness, and the type of steel as well as the amount of powder ignited. The expansion is greatest at the breech end of the barrel. This expansion moves down the barrel and decreases as the pressure drops during the bullet's travel down the barrel. If you picture a sausage-shaped balloon being squeezed from one end to the other, the balloon expands just ahead of the pressure. The same happens to a rifle barrel. Experiments have shown that if you fire a bullet into a water tank, the bullet diameter will be larger as you cut the barrel shorter because the barrel expands more at the breach end. It makes sense that heavier, thicker barrels will not expand as much as thinner barrels.

In addition to expanding in all directions, the barrel is violently twisted by the bullet being forced to rotate by the barrel's rifling. The force or torque required to start the bullet rotating causes a distortion on the barrel that increases as the bullet moves from the action to the muzzle. This rotation is counter to the rotation of the bullet and precedes the bullet down the barrel. These two phenomena explain the importance of a "free floated barrel" and not allowing the barrel of the rifle to come into contact with anything as the rifle is fired. The barrel must be allowed to move in its natural way. These forces place tremendous stress on the steel, in addition to the heat generated by the friction of the bullet and the hot

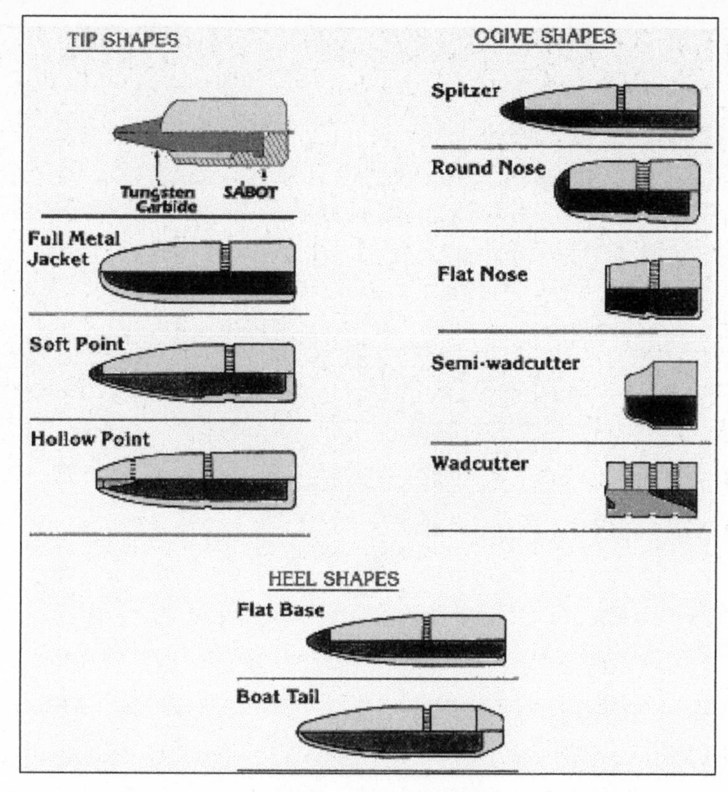


Different bullet shapes, whether on the tip or shank, are going to affect how a bullet flies and the terminal ballistics of the round.

gasses in the barrel. With enough stress, cracking and eventually barrel failure can occur. Any weak points in the grain texture of the steel are the most common places for these cracks to occur, which highlights the importance of a barrel being "stress relieved" during the manufacturing process.

Additionally, one may think that a rifle barrel is perfectly straight and that the strength of the steel over the length of the barrel would not allow the barrel to droop or sag. Wrong. Any material supported on only one end will sag slightly. As the bullet travels down the bore, the forces exerted on the barrel will try to straighten it. These forces create

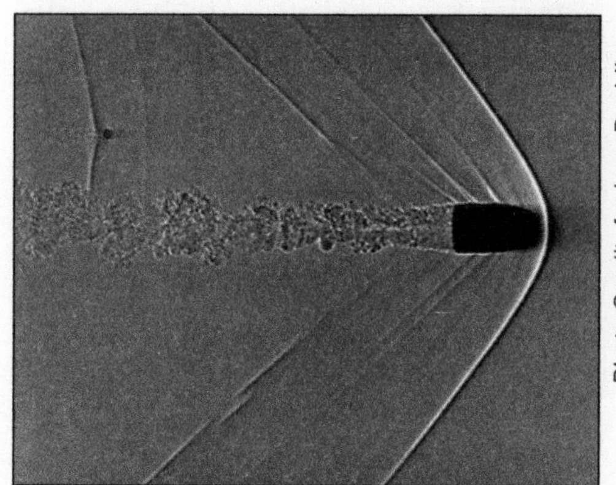

Photo Credit: Andrew Davidhazy

Shadowgraph of a supersonic bullet. Note the bow shockwave coming off the front of the bullet, and the air turbulence in the bullet's wake.

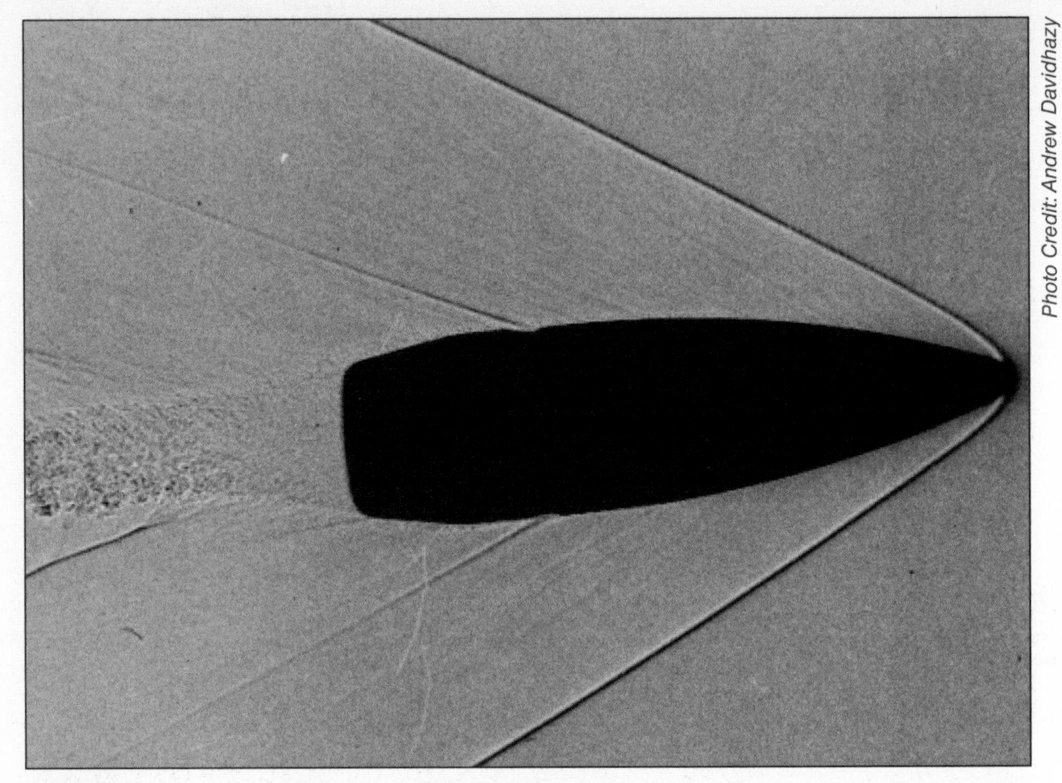


In this close up of a different round, you'll notice the boattail design, which will provide better accuracy over distance than the previous bullet's flatter base.

shock waves that cause vibrations or harmonics. All of these forces cause the barrel to have a whiplike motion. In most instances the barrel will not be perfectly straight as the bullet exits the muzzle. As you sight the rifle in, the optics will correct for this motion—as long as the motion remains the same. This again points out the need for the barrel to not be hindered from moving as it does naturally with every shot. Even improper use of a sling that is attached to the barrel or the stock coming into contact with the barrel can alter the bullet's point of impact. Additionally, this barrel motion will change with different ammunition, different powder loads, and different bullet weights, which again reinforces the saying that high accuracy is a result of high consistency.

A fair amount of technical intel, but the knowledge taken away is directly applicable in real-world shooting situations. Next time you throw a round off target while resting your barrel against a support, it will all become clear to you. Knowledge is power.

**Mural from Iraq. Lovely lady sporting the ever present AK 47, a widely used and reliable weapon.**

**Coauthor Glen Doherty in Iraq posing before demo shot (see next photo).**

Perhaps a bit more of an explosion than what happens in a rifle barrel. Blowing up some captured surface-to-air missles during the push to Baghdad.

Rifle Barrel Becomes Bugle for Musical Stunt

USING the barrel for a horn, an English musician can play bugle calls on a rifle. A trumpet mouthpiece is inserted into the muzzle and the bolt removed. The notes produced are shrill and piercing, but are said to be perfect in both tone and pitch. The originator of the idea is shown in the photograph reproduced at the left holding the novel instrument in the correct position for sounding a bugle call.

The musical rifle is played by blowing through it as if it were a bugle

Another way to place undue stress on a barrel . . . and everyone else around you!

## MOA (I'd Like Half an MOA, Please)

An important and commonly used acronym in a sniper's vernacular is MOA (minute of angle). What is MOA? Why is a rifle's potential MOA important?

Again, science, or in this case math, rears its ugly head. A minute of angle is a measurement of angular width. A complete circle is divided into 360 degrees. Each degree is further divided into 60 minutes and each minute has 60 seconds. Therefore one minute is one sixtieth of a degree. If you take that angular measurement from your shooting position it should form a triangle. One line is your direct line of sight (LOS) to the target, the second is a line rising from the muzzle at the one sixtieth of a degree angle, and the third line is the distance from the LOS and the sloping line.

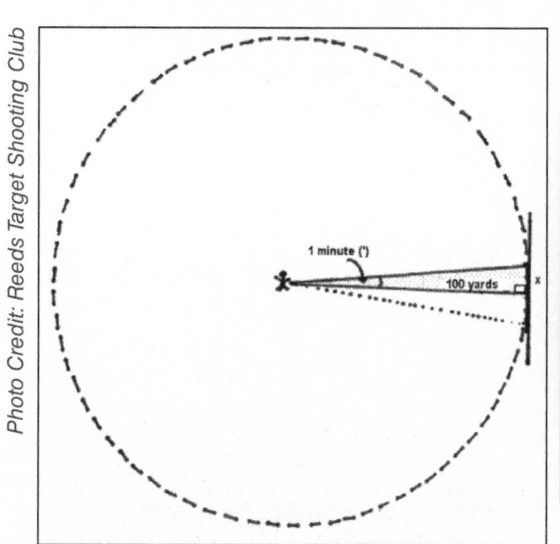


Photo Credit: Reeds Target Shooting Club

At 100 yards this distance is 1.0472 inches. In most shooting literature, 1 MOA is usually measured as 1 inch at 100 yards. The MOA increases linearly as you go further from the muzzle,i.e., 2 inches at 200 yards, 3 inches at 300 yards, and 10 inches at 1,000 yards.

Coauthor taking a break during some stress course work at Marc Halcon's American Shooting Center outside Alpine, California.

6

# What Makes a Good Sniper Rifle?

Rifles, especially precision rifles, are advertised to be capable of shooting a certain MOA. Most commercial hunting rifles will shoot 2 MOA or greater. A 2 MOA hunting rifle would, at its best, be capable of shooting a 3 to 5 shot group into a 2-inch circle at 100 yards or a 10-inch circle at 500 yards. If you are shooting an elk, which has a kill zone of 24 inches, this accuracy is acceptable as long as you can center the shot in the kill zone. If you are being asked to take down a hostile human target in a crowd, a 2 MOA weapon could very well be unacceptable and result in a miss or worse—a friendly getting hit. Most sniper rifles are at least 1 MOA capable, with the majority being .5 MOA capable. So to do the math, a .5 MOA sniper rifle will be capable of shooting a .5-inch group at 100 yards, or a 5-inch group at 1,000 yards. Now that's tight!

## One Rifle Isn't the Same as the Next

The history of rifle design and manufacture shows that rifles have evolved over time to be more effective at killing the intended target, be it game or humans. It is also a reality that human beings do not require any special large, high velocity round to

Photo Credit: Accuracy International

The Accuracy International's AS50 semiauto .50 BMG. The acronym designates this as a Browning Machine Gun .50 round, but it's also known as the "Badass Mackdaddy Gun."

be forced into taking a "dirt nap." A well placed .22-caliber air rifle has killed a child, and Robert Kennedy was gunned down with a .22-caliber handgun.

## Sniper Quality Rifles

So, what characteristic sets a sniper rifle apart from your typical hunting rifle?

### Accuracy

OK, OK, we know a sniper rifle has to be accurate, but how accurate is accurate enough? We noted that most hunting rifles are usually good for 2 to 3 MOA accuracy. Sniper rifles, on the other hand, are built to have a potential MOA accuracy of at least 1 MOA, which in reality is at the upper end of acceptable accuracy for a sniper-grade rifle. As with most things in life, you get what you pay for. A 1 MOA rifle will in all likelihood be less expensive than a more accurate rifle, because less expensive parts were used, it is built to looser tolerances, and has less "hand" build attention in its manufacture. A Remington Police Sniper Model 700 308 rifle is a good example of a 1 MOA rifle. This rifle will set you back around $1,000. With today's materials and knowledge, sniper rifles can be built to be capable of at least .5 MOA. The Remington M24 HS Precision Tactical rifles are good examples of .5 MOA sniper rifles. The cost of this grade rifle is about $2,000 to $2,500. There are some rifles that, with the right ammunition and right caliber, are capable of .375 to .25 MOA, but they will be significantly more expensive. Sniper rifles from GA Precision, Tactical Operations, and Accuracy

The AK is *definitely* not a sub 1 MOA weapon.

Photo Credit: Knights Armament.

The brand spanking new Knights Armament M110 SASS (semi automatic sniper system), a 7.62 match grade system with a lot of adaptability, is shown here with the digital desert camouflaged UNS LRLP (Universal Night Sight Long Range Low Profile), otherwise known as the PVS-26—an amazing setup for night operations.

International are good examples of this grade rifle, but the cost is $4,500 to $6,000 for the rifle alone. Add another $2,000 to $3,000 for good quality glass, and you are out $8,000 to $10,000 for your new toy. Let's look at some grade components.

### The Barrel

Each manufacturer will, of course, claim that their barrel is the best and tell you why their way of manufacturing results in a high degree of accuracy. Sniper-grade barrels are constructed from both chromoly and stainless steel. Stainless is more expensive and may resist rust and wear better than chromoly.

In general, most sniper-grade barrels are thicker than commercial hunting rifle barrels. The extra thickness is beneficial to the barrel accuracy because it is stiffer, and therefore has less whip and motion during bullet transit. The additional thickness also makes the rifle heavier, especially toward the front of the rifle. This makes the shooter feel less recoil, again adding to the accuracy and shoot-ability. The added forward weight also decreases the muzzle jump, which helps the shooter stay on target and if necessary rapidly put another accurate round down-range. The added thickness will also help deal with the heat of repeated fire. The added weight

Photo Credit: Barrett

This Barrett Model 95 is a shorter, lighter version of their bolt action .50 cal, the Model 99, and sports a (relatively for a .50) short 29" barrel.

is, of course, a concern to the sniper having to carry the rifle into the field, but the weight translates to improved accuracy and in most cases is an acceptable trade-off. Besides being thicker, most sniper-grade barrels are also longer than hunting rifles. The length of the barrel should be tailored to the caliber and the powder load. Again, you want the barrel to be long enough to maximize the full benefit of the gas developed by the powder. The shorter the barrel, the less the muzzle velocity; too long a barrel with an improper powder load could also reduce pressure, which could result in a lower muzzle velocity. Shorter barrels are stiffer, but the accuracy may suffer, especially at longer ranges, again because of the decreased velocity. Shorter barrels are easier to maneuver, but most snipers will not be doing house clearing with their rifle. But shooting offhand or in non-supported positions is easier with a shorter, lighter rifle. *Most* sniper rifles, especially for eight hundred to one-thousand plus shooting will have 24- to 27-inch barrels. Additionally, a shorter barrel will have a louder muzzle blast, which if unsuppressed could be a concern to a sniper trying to hide his location from the enemy.

### Rifling Explained

Rifling varies from manufacturer to manufacturer. The cheapest method

**The authors would love to find this complete kit under the Christmas tree some year.**

*Photo Credit: written and illustrated by Robert A. Rinker*

**A mandatory read for anyone interested in the science and art of firearm ballistics.**

is hammer forging. Some very accurate rifles—the Remington 40 xb, the Remington M24, and Sig Sauer rifles—use this method to "cut" the barrel rifling. Custom barrel manufacturers, such as Bartlien, Schneider, Lawton, and Krieger, employ cut rifling exclusively. Modern technology tolerances for the bore and the rifling can be as small as .0002 inches, which, as those of us that managed to stay awake in fourth grade math know, is very tight.

All of these rifling processes will leave some imperfections that will need to be removed with some kind of lapping process. Lapping will make a mediocre barrel better, but not great, so what really matters is quality in the entire barrel building process. You may wonder why it is important to remove theses imperfections from the barrel: After all,

**More art from Iraq**

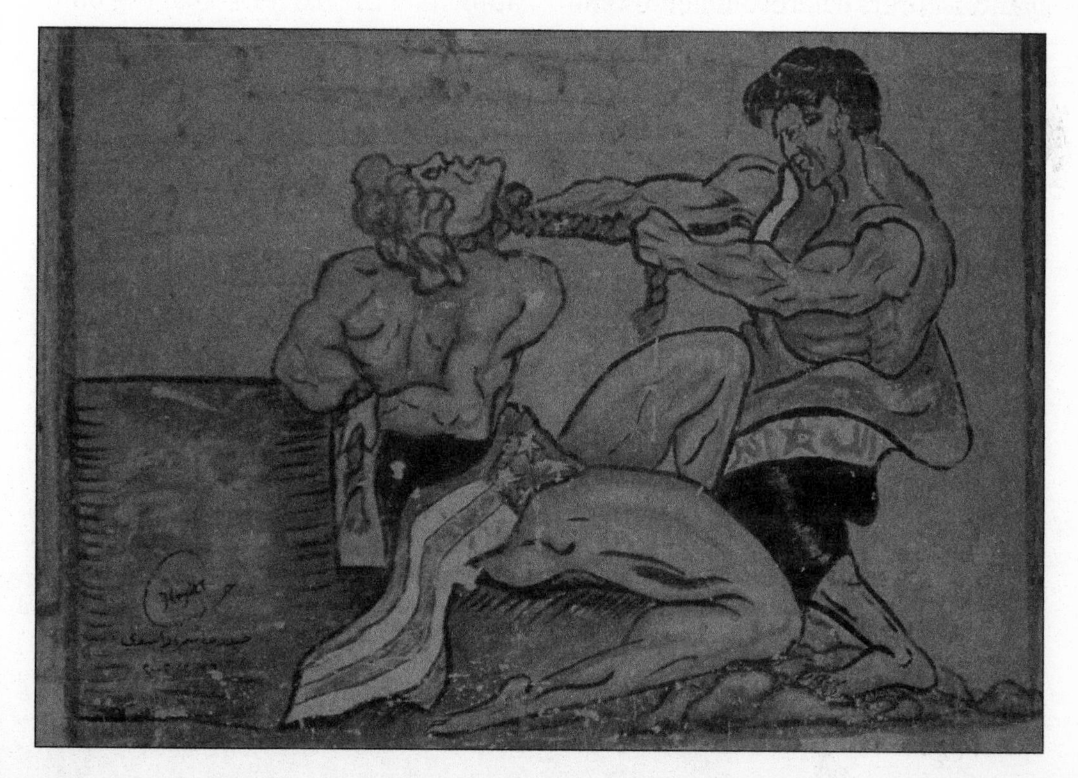

the bullet is being pushed down the barrel in most instances at two to three times the speed of sound, so how can little imperfections in the barrel degrade the barrel's accuracy? These little imperfections interfere with the bullet *smoothly* moving down the barrel. They also strip away parts of the bullet's copper jacket, so the bullet leaves the barrel not perfectly symmetrical, which interferes with its gyroscopic stability during its flight to the target. The copper left behind in the barrel can build up, adding to the "roughness" of the barrel's bore and the fouling of the barrel during subsequent shots. An unstable bullet is an inaccurate bullet. Most manufacturers will recommend some kind of "break-in " process that is designed to help smooth out these imperfections. The break-in process is time consuming, usually requiring a series of shots, cleaning the barrel between single shots, double shots, and finally three- or five-shot groups. The theory behind this is that as you shoot rounds and then clean the barrel, these small imperfections are shot smooth. Additionally there are some abrasive bullets offered by David Tubbs, a multitime long-distance rifle champion, that are supposed to smooth the barrel quickly and more efficiently than the laborious break-in period. If the barrel manufacturer recommends a certain break-in process, it may be best to follow their routine so that they will honor any warranty. Lastly, there are some products that can be applied to the barrel to coat the bore. Again the barrel or rifle manufacturer can recommend these products, based on their experience and their product.

Photo Credit: Curtis Prejean

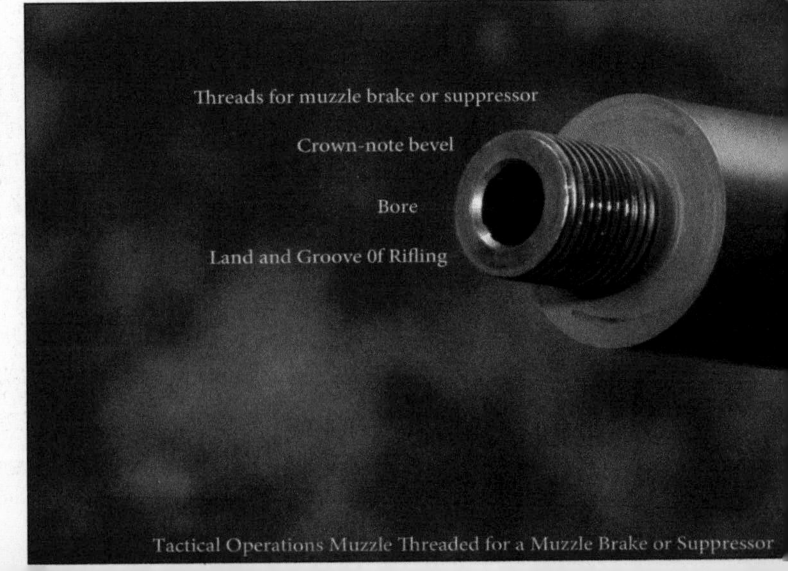


**Detailed photo of the pointy end of the spear**

**Twist, Crown, and Jump**

The twist rate, as noted earlier, is the rate at which the helical or spiral pattern turns over a certain length of the barrel. A twist rate of 1 to 10 means that the rifling completes one complete revolution every 10 inches. You remember that this twist causes the bullet to rotate as it moves down the barrel, which gives the bullet gyroscopic stability during flight. Having the proper twist is very important to accuracy. Again, an unstable bullet is an inaccurate bullet. The proper twist rate can be determined with ballistic formulas that are too detailed to get into in this chapter. If you are interested in the math/science you should check out *Understanding Firearm Ballistics* by Robert A. Rinker. Rifle barrel manufacturers will usually match the twist rate to the intended ammunition for the sniper rifle. Most sniper-grade 308s, 300 Win Mag, and 338 Lapua rifles have a 1 to 10 twist. As a general rule, the heavier bullet will require a faster twist, so if you know that you will only be shooting a 300 grain 338 bullet out of your 338 Lapua sniper rifle, you may want a 1 to 9 or 1 to 5 twist rate. Again, custom built rifles can have this kind of "tweaking," but it will cost more money.

The crown on a rifle barrel refers to a bevel cut at the muzzle. The crown of the muzzle is the last part of the rifle that will come in contact with the bullet as it leaves the rifle. Any imperfections, indentations, or nicks in the bore at the end of the barrel will impart some instability to the bullet.

San Diego Padres' incredible third baseman Kevin Kouzmanoff gets some with the author's .300 Win Mag. Notice the rear accessory pack and speed loader.

SEAL heads North from Baghdad manning the .50.

Quality sniper rifle manufacturers will bevel the end of the barrel to ensure that the bore is free from any imperfections. Usually this machining is beveled in so the crown is recessed from the end of the barrel to help protect the crown or bore from being hit or dinged during operations. If the barrel is threaded to take a suppressor, there is usually a crown protector that can be put on the end of the barrel. If the crown is damaged you will see a drop in accuracy. A quality gunsmith can recrown the barrel, so don't despair if you drop yours muzzle first off the top of a ten story building.

The jump or bullet jump refers to the distance from the end of the chamber to the beginning of the rifling. As noted earlier, most rifle manufacturers cut the chambers to match standard cartridge dimensions. Most sniper rifles will have this as the standard chamber. Why? Because the rifle has to shoot all brands of ammunition issued to the sniper. There will be slight deviations in cartridge length from manufacturer to manufacturer and from lot to lot (a "lot" of ammunition is a quantity of ammunition assembled by one producer under similar conditions and is expected to produce similar results on target). Most snipers, especially in the military, do not have the option of hand loading all of their rounds—they shoot what they are given, so their rifles must be capable of shooting all ammo. If you can hand load and you want to further customize your rifle, you can have the barrel builder shorten the jump to tighter tolerances, or seat your bullets out slightly longer during the reloading process. Some reloaders try to have the ogive (curved sides) of the bullet actually touch the lands and grooves of the barrel. As always, there are trade-offs for doing this, and doing this may only give a slight increase in the accuracy of the rifle. Bench rest shooters are more likely to benefit from this than an active sniper. You should also note that having your chamber cut to precisely match the cartridge's overall length so that there is minimum-to-no bullet jump is for lead core bullets. You should not do this if you are going to use solids, since the solid bullets do not conform to the barrel as easily as lead core bullets. You could get into problems with very high pressures that could damage your rifle or you!

### Barrel Life

No, not you—how long will a sniper-grade barrel "live" or last? That depends on a lot of factors. Stainless steel is a bit more wear resistant than chromoly, but the three biggest factors affecting barrel life and, in turn, long-term barrel accuracy are the velocity of the round, the number of rounds through the barrel, and barrel care/maintenance. High velocity rounds (greater than 3500 to 4000 fps) will wear a barrel much quicker than slower cartridges. Again there is a trade-off in terms of wanting a flatter trajectory, less wind interference, higher energy, and being willing to buy a new barrel. The number of rounds through the barrel is important to keep track of, because over time you will see a decrease in accuracy as your barrel wears from shooting. Maintenance of your barrel—good cleaning, not scratching the bore, and keeping the crown from getting damaged—will make the barrel last a long time. Want an estimate for a sniper rifle barrel life? A good sniper-grade 308 barrel that's well maintained and never shot until it's red hot should last 8,000 to 10,000 rounds, unless you are pushing the loads as hard as you can. (That doesn't sound good!)

## Don't Touch Me!

Accuracy is based on consistency—consistency in everything the rifle does as well as the human factors involved with making a good shot. The barrel should not come into contact with anything when the rifle is fired. This includes the stock. This requirement is known as "free-floating the barrel." A quality sniper rifle will have the stock and the rifle barrel set so that the barrel does not come into contact with any part of the stock. The same cannot be said about all "hunting" rifles. You can check to see if the barrel is free-floated by running a sheet of paper folded in half along the barrel that is above the fore end of the stock. Some .50-caliber builders will eliminate this problem by not having a fore end to the stock. This principle now appears in a lot of AR-style rifles where the fore end is a "free-floated tube" that does not touch the barrel at all.

Photo Credit: www.tikka.fi

Test for a free-floating barrel

## Triggers

There are two basic types of triggers: single stage and two stage. In a single-stage trigger, there is virtually no uptake in the movement of the trigger as the shooter applies pressure to break the shot. The pressure required to cause the trigger to "break" or fire the rifle can be altered with some custom trigger work or in some cases the trigger may actually be adjustable. You want the trigger pull to be smooth and consistent. A trigger job can smooth parts in the trigger assembly to improve trigger pull. Single-stage triggers are the most frequent type of trigger used in American made sniper/tactical rifles. Most tactical/sniper rifles have a trigger pull of 2.5 to 3 pounds.

Photo Credit: Ammoland.com

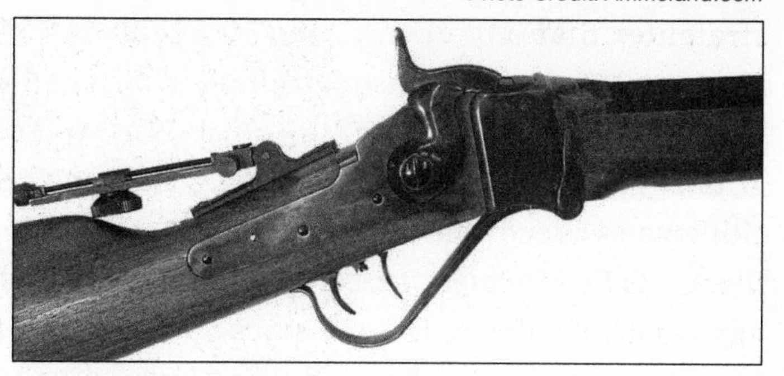

**Here's an old school double set trigger on an Sharps Model 1874 sporting rifle. The rear trigger is pulled first, which then allows the front trigger to fire the round.**

You don't want a trigger that takes a tremendous amount of effort to break, but you also don't want a trigger that will break if you slam the bolt closed or that will break with the slightest pressure before you are sure of your target and have the green light to engage the target. More weight will add stress to the sniper during the natural respiratory pause of breathing that is generally recommended as the time to break your shot. A trigger can be adjusted to less than 2.5 pounds (all the way down to 8 ounces) but too light a trigger can resutlt in accidental discharges. Remember, once you squeeze the trigger there is no calling the bullet back!

Two-stage triggers are called this because there are two stages of engaging the trigger mechanism. The first stage is pulling up the slack in the trigger movement, which is fairly light. In the second stage, the required trigger pressure is higher but remains constant until the trigger breaks. Two-stage triggers are standard on Accuracy International rifles and other European rifles. It should be noted that most rifle actions will only work with one style trigger. You usually cannot switch out a single-stage trigger assembly for a two-stage trigger, even if the two-stage trigger is what you prefer.

Remington triggers are easily tuned and they offer adjustable triggers on some of their higher end rifles. Aftermarket triggers from Timmey and Jewel are available for rifles built with a Remington Model 700 action (Tactical Operations and HS Precision).

### Sniper Rifle Stocks

"This is my rifle. There are many like it, but this one is mine.

My rifle is my best friend. It is my life. I must master it as I must master my life.

My rifle, without me, is useless. Without my rifle I am useless. I must fire my rifle true. I must shoot straighter than my enemy who is trying to kill me. I must shoot him before he shoots me . . ."

This part of the Marine Corp's "Rifleman's Creed," made famous by the movie *Full Metal Jacket*, certainly applies to all snipers. It is the stock that lets a shooter hold, nestle, and fit a rifle to his body, becoming one with the rifle. To accurately place rounds down range the sniper must be able to consistently position the rifle with his body and support it the same way on every shot. Put another way, a sniper needs a rifle that "hits and fits." In order for this to be accomplished, most sniper-grade rifles have adjustable stocks that can be changed to fit the sniper. Like a finely tailored suit, the rifle needs to be fitted to the shooter. The two parts of the rifle stock that are adjustable are known as the length of pull and the cheek well. The length of pull is the distance from the place you place your firing hand to the butt of the rifle stock. A good quality sniper rifle stock can be lengthened or shortened as needed by the sniper. Why would this distance need to be changed? Shooting with full body armor will add size to the sniper as opposed to shooting in a plain uniform. Shooting in the dead of winter will usually have the shooter dressed in warm, bulkier clothes as opposed to the dog days of summer. Additionally, different shooting positions will change the distance the shooter needs to get proper eye relief with the optics. The cheek well is the top edge of the rear portion of the stock, where the sniper rests his cheek during shooting. This, too, needs to be adjustable, depending on the shooters face/head structure and the height of the optics on the rifle. The height of the cheek well should be adjustable so the sniper can consistently place his cheek on the rifle in the same place every time. It should be adjusted so that the sniper can almost "fall asleep" on the stock, in such a way as to

**The J. Allen Enterprises JAE-700 RSA Precision rifle stock for the Remington 700 short action is part of a new generation of fantastic stocks on the market.**

*Photo Credit: www.jallenenterprises.com*

Photo Credit: Curtis Prejean, courtesy of McMillan

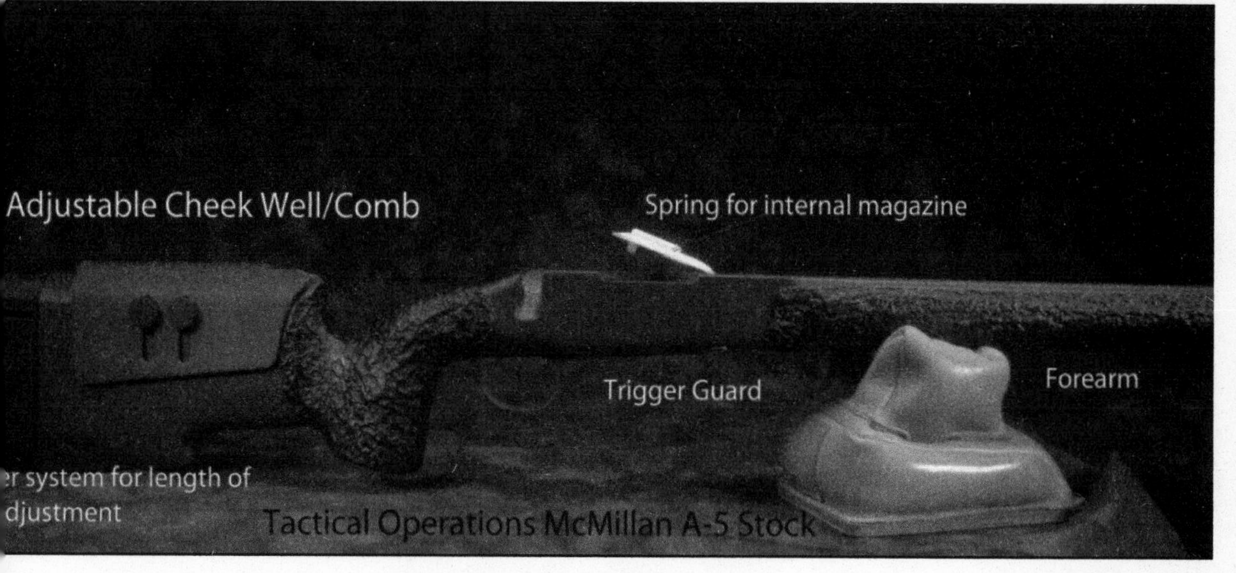


**Tactical Ops stock from McMillan. Nice.**

not be straining his neck or head to contort himself to the rifle. Despite what Hollywood shows, not all sniper missions entail shooting hundreds of rounds; many times there are long periods of observation, calling in air strikes or waiting for the target to present a shot. The rifle must fit the sniper, not the sniper fit the rifle, so that the sniper can comfortably position himself and his rifle to observe the enemy for long periods of time.

Today most sniper-grade rifles are made of synthetic/fiberglass materials as opposed to wood. The fiberglass stocks from companies such as McMillan, HS Precision, JAE, and Mannard are very strong, rigid, and do well in a wide range of operating conditions. Many of these synthetic/fiberglass stocks have aluminum pillars for holding the action in the stock, but the action should be "bedded" into the stock with additional epoxy or fiberglass so that the action of the rifle is snugly nestled into the stock.

There is also a trend toward an all-metal or modular stock, or chassis, that does not require any bedding. Accuracy International Rifles have an aluminum chassis that simply requires that the rifle action be secured to the chassis with the appropriate screws with the proper amount of torque. McCree and the new Remington MSR Sniper rifle use modular aluminum stocks.

Sniper stocks, as compared to hunting rifles, will have areas usually on the stock forearm that are designed to hold a variety of sniper accessories, such as lights, lasers, night vision optics, and bipods.

### The Bolt and the Action

The bolt and the action should be machined so that the bolt's locking lugs perfectly match the action and should be perfectly aligned with the chamber and bore of the rifle. The bolt face that comes in contact with the base of the cartridge should be perfectly flat so that the cartridge cannot be bent, warped, or twisted in the chamber. Any misalignment will result in decreased accuracy. The firing pin that is housed in the bolt should be set to strike the cartridge with enough force to detonate the primer but not so much as to cause excess vibration in the rifle. "Lock time" is the time that passes between your trigger finger releasing the trigger mechanism and the firing pin striking the primer. Lock time should be as fast as possible so that the rifle has no time to move between the instant you actually activate the firing pin and resulting ignition of the powder. Lock times vary from 0.0022 to 0.0057 of a second. You can have the lock time adjusted with different springs and also by replacing the steel firing pin with one made of titanium. Again, you get what you pay for.

If you hear that that bolt has a "tactical" knob, all this means is that the bolt operation lever is a little

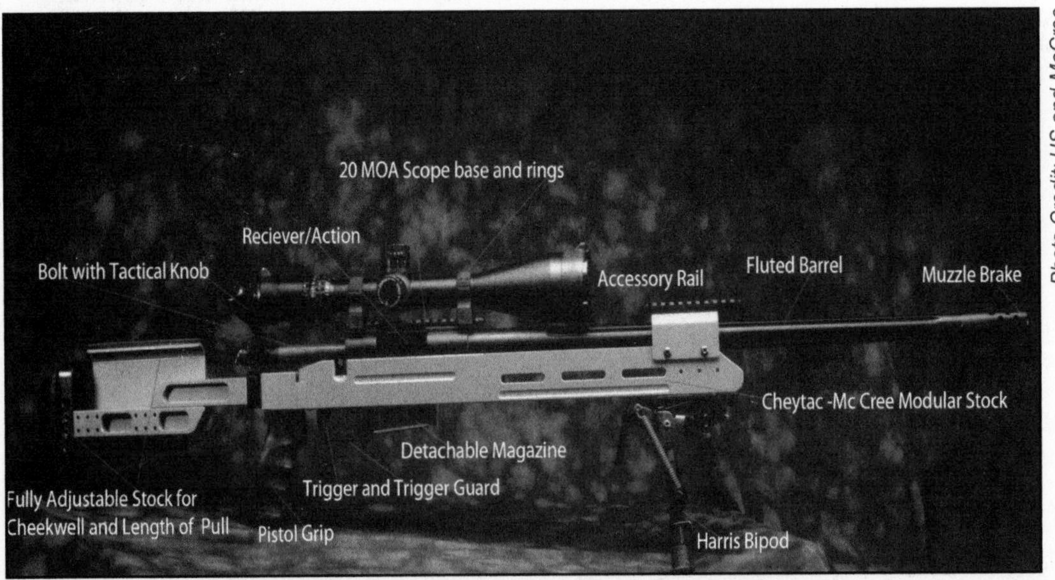


Photo Credit: HS and McCree

Detailed view of an HS.308 Witra McCree aluminum stock

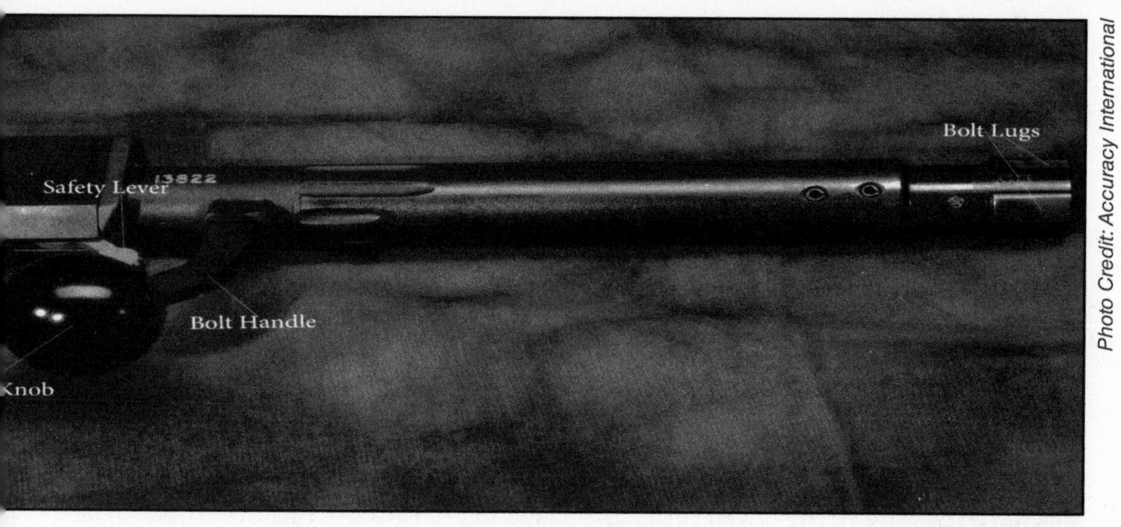


Photo Credit: Accuracy International

Accuracy International bolt

longer and may have a bigger knob on it, compared to a nontactical rifle, since a sniper may be using gloves. However, the biggest reason for a tactical bolt lever or knob is that under the stress of combat, your fine motor skills decline at least 40 percent, so having a larger knob to grab and manipulate may be a benefit when you are being shot at.

## Muzzle Brakes

Depending on the caliber or the shooters preference, the barrel may be fitted with a muzzle brake. A muzzle brake will definitely reduce

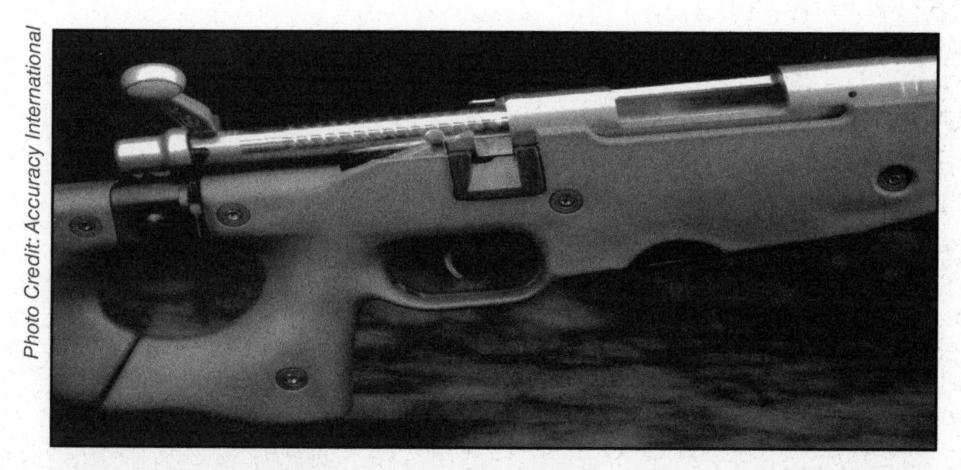

Photo Credit: Accuracy International

The Remington Mil-stock R bolt and action in an Accuracy International aluminum chassis. Note the machine work.

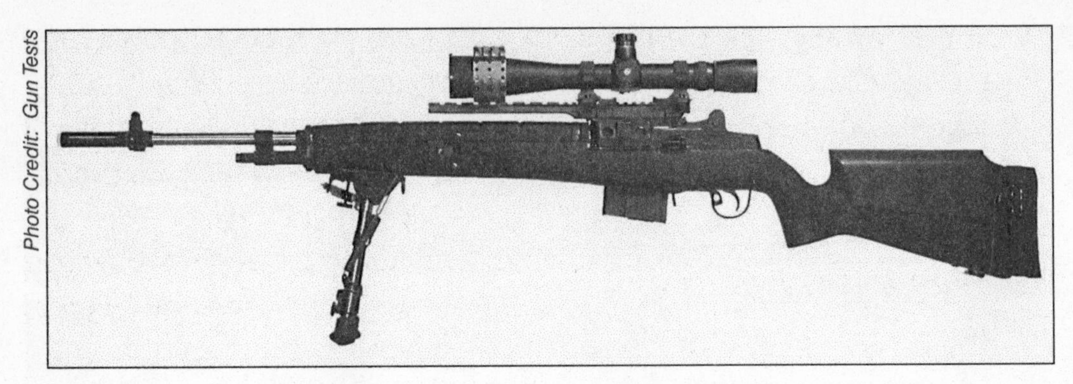


Sweet McMillan 3A stock on a Fulton Armory M14 rifle

the felt recoil but will increase the sound of the rifle firing. Most muzzle brakes will on average reduce recoil about 30 percent. Smaller caliber rifles (.223, 6.5 mm, .308 Win) may not be fitted with a muzzle brake, since the recoil of those calibers is not an issue for most shooters. However, the muzzle brake's reduction in recoil and barrel jump do help keep the rifle on target and allow the sniper to put additional shots down range more quickly. For larger cartridges (.300 Win Mag, .338 Lapua, .408 CheyTac, and .50 BMG), muzzle brakes are much more common as the recoil without the brake will be much more difficult to manage. A suppressor is in some snipers' minds "the ultimate muzzle brake." Suppressors will reduce felt recoil and will drastically reduce the sound of the rifle firing—a great benefit for a sniper in hiding from the enemy. If you plan to shoot with a suppressor, you should be sure to zero the rifle

One of the many different styles of muzzle brakes

with and without the suppressor as the addition of a suppressor will change the weapon's zero to some degree.

## A Summary of What Makes a Great Sniper Rifle

1. A match-grade, free floated barrel. This will be thicker than regular hunting rifle barrels and manufactured with tighter tolerances. Hopefully this will give the barrel the potential for sub 1 MOA accuracy. The barrel should have been stress relieved during manufacturing and should be lapped to reduce as much of the defects associated with the boring and rifling process as possible. BREAK IN YOUR BARREL ACCORDING TO MANUFACTURERS' SPECIFICATIONS! If you don't, you'll sacrifice barrel life and accuracy.
2. An action mated to the weapon constructed with the highest standards.
3. If shooting a large-caliber weapon, ensure you are using a muzzle brake or suppressor. Your shoulder will appreciate it if putting a lot of rounds down range, and if you are shooting from a concealed position, the suppressor will ensure you STAY concealed.
4. An adjustable stock, quality bipod, and the best optics you can afford.

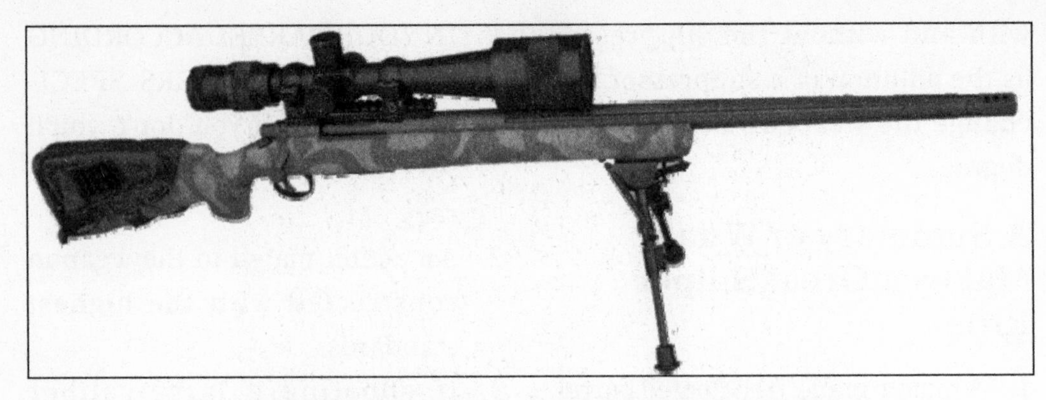

This Remington 700 has a Speed-lock titanium firing pin kit with the U.S. Optics SN3 Variable power scope and a Harris Bipod, and was built by ARM U.S.A's Darkhorse Gunworks.

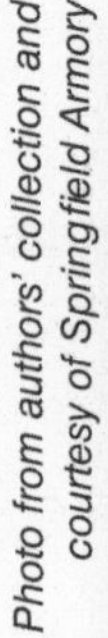

Coauthor getting the feel for Springfield Armory's re-release of the M21 Sniper System. Basically an M1A Supermatch with a custom Douglas barrel and an adjustable cheek piece stock. I love this rifle.

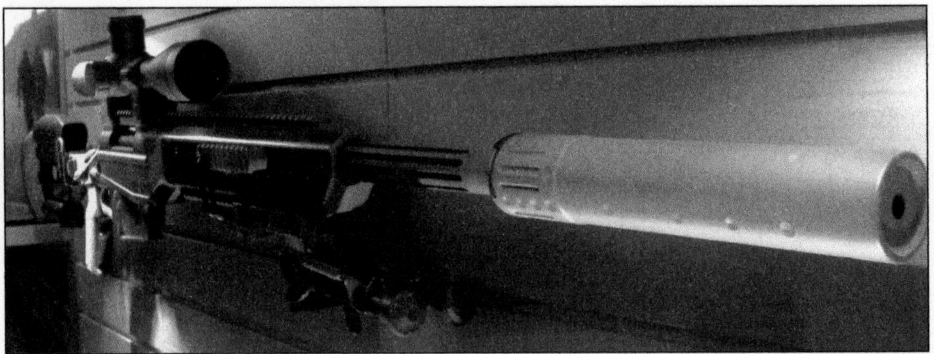

Photo from authors' collection and courtesy of Ashbury International

Ashbury International's Asymmetric Warrior ASW338LM Precision Sniper Rifle System. This sweet weapon uses Rock Creek Pinnacle barrels and comes equipped with an integral muzzle brake and sound suppressor system. This cutting edge U.S. company is pushing the technological envelope with their weapons systems. Expect to see more from them.

Photo courtesy of Curtis Prejean.

**Example of a "sniper grade rifle," the Accuracy International 338 Lapua Magnum**

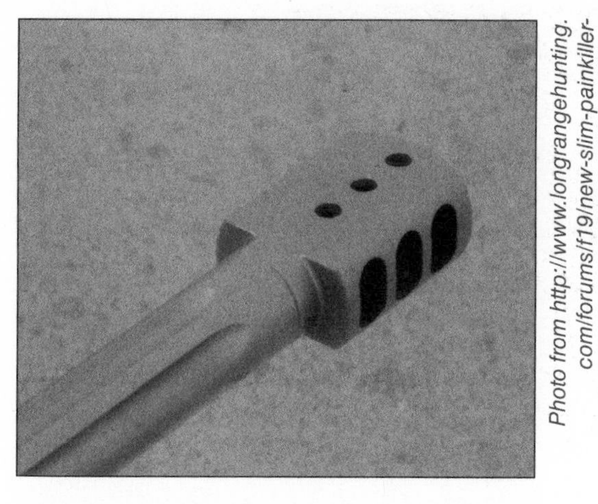

Photo from http://www.longrangehunting.com/forums/f19/new-slim-painkiller-muzzle-brake-aps-33319/

**The Allen Precision Shooting "Painkiller" Muzzle Brake**

Photo from Curtis Prejean and courtesy of Accuracy International

**Accuracy International Muzzle Brake**

7

# Beginner Ballistics, Effective Range, and Potential Applications

Humans do not do well with bullet holes. Granted there are some instances in which a human being can sustain multiple gunshot wounds and continue to function for a time depending on the extent of the injury, the location of the injury, and access to appropriate medical care. There are many examples of a drug- or adrenaline-fueled combatant requiring multiple rounds to stop him, unless you can place a shot to his brain. In fact, it is a well-known fact that even with a shot to the heart, an appropriately motivated enemy can live for five to eight seconds—more than enough time to shoot you in a close quarter gun battle. So given these facts of human frailty, why not use any caliber? Two main reasons: a) the sniper needs the right tool, such as rifle and caliber combination, to perform his mission, and b) the sniper needs an accurate caliber/bullet combination to match his sniper rifle. Accurate caliber you say? Yes, not all calibers and bullets are created equal.

There are three main types of ballistics. Internal ballistics relate to what is happening in the rifle as the bullet leaves the neck of the cartridge. External ballistics relate to what happens to the bullet as it flies to the intended target. Terminal ballistics relate to what happens to the bullet when it hits the intended target. Sniper cartridges, like sniper rifles, must be accurate out to the range that the sniper has to deliver a shot. Additionally, the cartridge/bullet combination must have the proper terminal ballistics, such that the velocity and energy to perform the intended job. Different targets will obviously require different bullets. A bullet's ability to take out a target depends on the energy or force needed to penetrate and destroy the target. A head shot at 600 yards will require a certain amount of bullet energy to penetrate the skull. If the enemy is behind a glass windshield, a door, or a cinder block wall, obviously the bullet will need more energy to penetrate the cover and still produce a kill.

Over the years the military has done intensive studies on what caliber bullet is best for particular sniper operations. Six calibers are most frequently used in our current environment. All of these cartridges are very accurate, and all have their advantages and disadvantages and will to some degree be chosen

**Knights Armament M110 is replacing Remington's M24 as of 2010 as the Army's primary Sniper System. It is shown here with the PV S26 Universal Night Sight Short Range (UNS SR).**

*Photo Credit: Knights Armament*

Photo Credit: Curtis Prejean

**Calibers commonly utilized for sniper rifles**

depending on the sniper mission. The six calibers are: .223 Remington, .308 Winchester, .300 Winchester Magnum, .338 Lapua, .408 CheyTac, and the .50 BMG. Other, less-popular calibers include the 7mm Magnum and the .419 Barrett. Out of these six calibers, the .308 Winchester and .300 Win Mag are probably the most common. They are plentiful, relatively inexpensive, useful in a wide range of missions, and very accurate out to their maximum effective ranges. The .338 Lapau Magnum is seeing more use, but will most likely only be found in the hands of a Special Operations unit. The .338 ammo is also much more expensive (a box of 20 costs $120 on the civilian market).

The .408 CheyTac and the .50 BMG both do extremely well at long range. By long range, we are talking a mile to a mile and half—1,600 to 2,400 yards. They are also used for taking out hard targets—vehicles, radars, communication equipment, engines, or other "hard" objects. Several snipers from multiple allied countries have set new records in the global war on terror with confirmed kills at ranges beyond two thousand yards.

## Ballistic Coefficient

The Exbal ballistic charts in Appendix A show information for the

168-grain Sierra Match King bullet and the 167-grain Lapua Scenar bullet. Both bullets are hollow point boat tail (HPBT) in design, but the Lapua Scenar has a better ballistic coefficient (BC), which is a measure of how well a bullet can overcome air resistance and keep its flight speed. A perfect BC is 1. The larger the BC, the better the flight characteristics. The bullet's ability to keep as much of its muzzle velocity during its flight is an important factor in both the trajectory and the killing effectiveness of the bullet. The characteristics that affect the bullet's BC are its weight, diameter, shape, and drag coefficient.

If you look at both the Match King and Scenar ballistic charts you will note that the Scenar bullet has a slightly better BC: 0.47 compared to 0.458. Not much, but note the velocities, retained energies, and the bullet drop. A bigger BC is better.

## The .223 Remington

The .223 Remington was originally designed for use in the military's AR 15/16 in 1957. It is very accurate and relatively flat ,shooting out to about 600 to 800 yards. At about 800 yards, the 77-grain bullet will go subsonic and become unstable. Most believe the effective range for this cartridge is about 600 yards for sniper applications if the target is a soft target. The round does not have the needed energy to penetrate ceramic plates in body armor. There have also been instances in Iraq and Afghanistan where the .223 does not have the energy to effectively eliminate a combatant that is high on drugs or has minimal improvised armor. Most sniper teams will have an AR-style weapon as a backup for retreating if discovered, since the AR, with its semi-or full-auto operation and thirty-round magazines, will offer a weapon that is easier to wield in closer combat, with easier target acquisition at ranges inside 400 yards.

You should be aware that bullets, unless otherwise designed, are supersonic when they leave the muzzle. There are special applications for subsonic ammunition. Ballistic science has shown that when bullets transit from supersonic speed to subsonic, they become unstable in flight and thereby the accuracy greatly decreases. The speed of sound will vary depending on many environmental factors. At sea level in 68°F dry air it is 1,125 fps. For you metric people, the speed of sound in the same circumstances is 343 meters per second. (Other measurements of the speed of sound: 1,236 kilometers per hour, 768 miles per hour.)

So when you are consulting a ballistic table or using a ballistic computer, note the distance at which the bullet drops below 1,125 fps. After this distance, your chances of making an accurate hit on target are drastically less. Again, in order for the bullet to take out the target, it will also need enough energy, which is dependent on the speed it is flying and its mass. As anyone who went to sniper school, or even took high school physics, will recall, kinetic energy is given by the formula $\frac{1}{2} mV^2$ where "m" is the mass (not weight) and "v" is the velocity.

## The .308 Winchester

The .308 Winchester was developed at the end of World War II. The U.S. Ordnance Corps began looking for a smaller cartridge to replace the .30-06 Springfield. The cartridge went through numerous changes and was eventually adopted by NATO as the 7.62 NATO at the end of 1953. The United States adopted it for use in the M14 and the M60 machine gun in 1957. Winchester introduced the cartridge to the public and chambered a Model 70 rifle for the cartridge, hence the name .308 Winchester.

The .308 Win is an extremely accurate round and is thought by some to be the most inherently accurate .30-caliber cartridge ever produced. It is probably the most common cartridge for most sniper rifles, especially if the missions don't require shots past one thousand yards

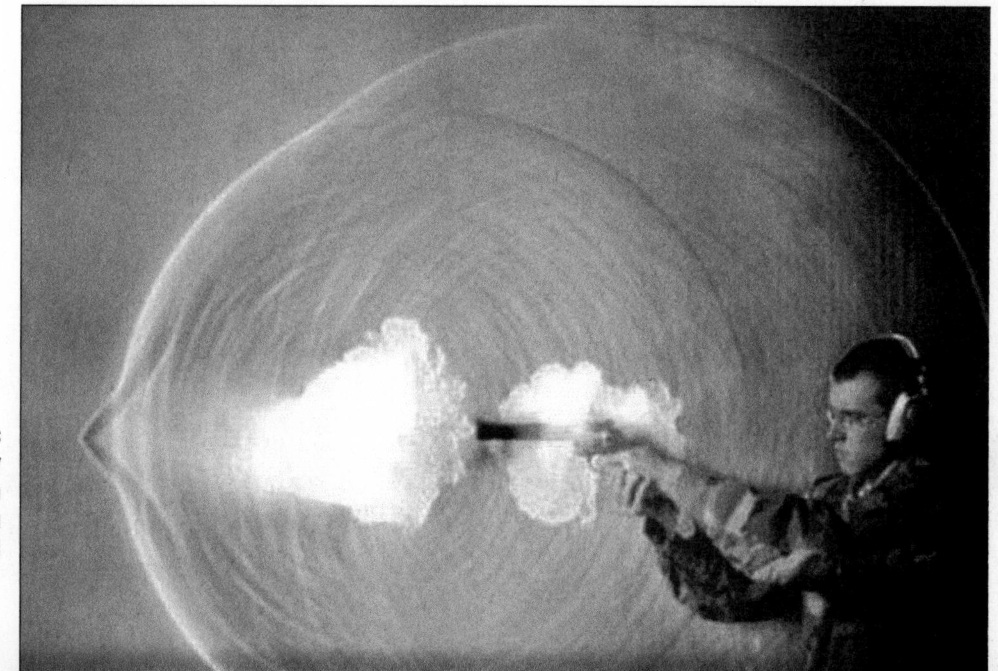

Sound waves as they emanate from a discharging pistol

or hard target penetration at extended ranges. There is ceramic plate body armor that will stop a .308, but Kevlar is not a problem for the round. The most common bullet is the 168 hollow point boat tail, usually abbreviated as HPBT. The "hollow point" is not actually a true hollow point as you would see in some pistol personal defense ammunition but is a result of the copper jacketing process.

The .308 Win is a very accurate and versatile round. Most professionals would say that its effective range is about 800 yards, but snipers have made shots on soft targets out to a 1,000 yards. If you consult the ballistic charts in Appendix A you will note that both the Scenar and the Match King bullets are still supersonic at 1,000 yards, 1,201 fps and 1,142 fps respectively. The Scenar goes subsonic at just under 1,100 yards. Both bullets retain energy that is about one-third of a .44 Magnum at point-blank range!

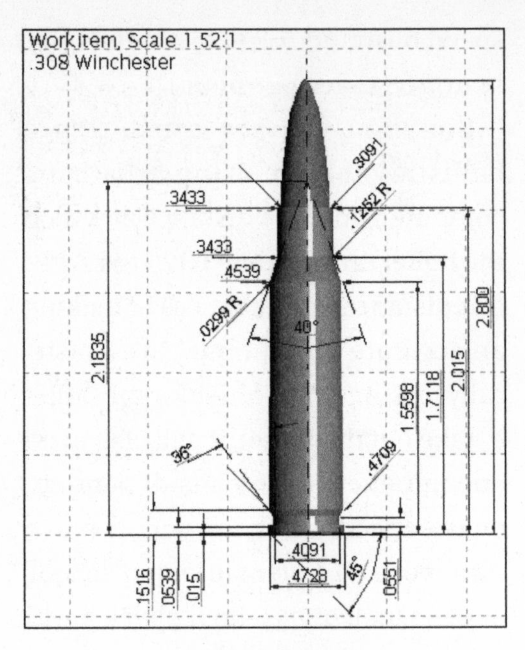


The .308 Winchester

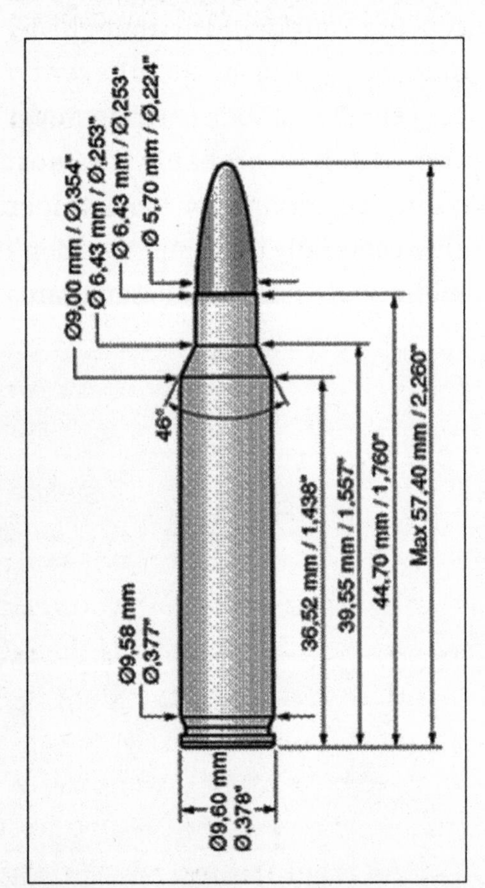


Exterior measurements of the .223 Remington

## The .300 Winchester Magnum (.300 Win Mag)

The .300 Winchester Magnum, usually called the .300 Win Mag, was introduced by Winchester in 1963. The .300 Win Mag is used in sniper

Some sweet 7.62mm (.308) Springfields on display at the SHOT show in Las Vegas

*Photo from author's collection and courtesy of Springfield Armory.*

rifles when there is a possibility of potential targets between 1,000 and 1,200 to 1,300 yards. (More of a guideline than a rule . . . no reason not to shoot a close target with a .300.) The cartridge for sniping is usually a 190 Grain HPBT bullet. The bullet goes subsonic somewhere between 1,350 and 1,400 yards. At this range the bullet still retains about the same energy as the .308 had at 1,000 yards. The .300 Win Mag is also a very accurate cartridge but does have more recoil than the .308 Win. Most .300 Win Mag sniper rifles will therefore be fitted with a muzzle brake.

The Remington M24 sniper rifle can be chambered for either the .308 or .300 Win Mag. All of the top sniper rifle manufacturers have the .300 Win Mag as an option for those shooters requiring a rifle that is accurate out to the 1,300-yard mark.

## The .338 Lapua Magnum

The .338 Lapua Magnum was originally developed in the 1980s for the Navy SEALs as a long range sniper rifle by Jerry Haskins of Research Armament Industries and Jim Bell and Boots Obermeyer. At this time the cartridge was called the .338/.416 since its case was based on a .416 Rigby cartridge. Lapua developed the .338/.416 further by strengthening the case and getting rid of the original belted head design of the .416 Rigby. Therefore, in reality,

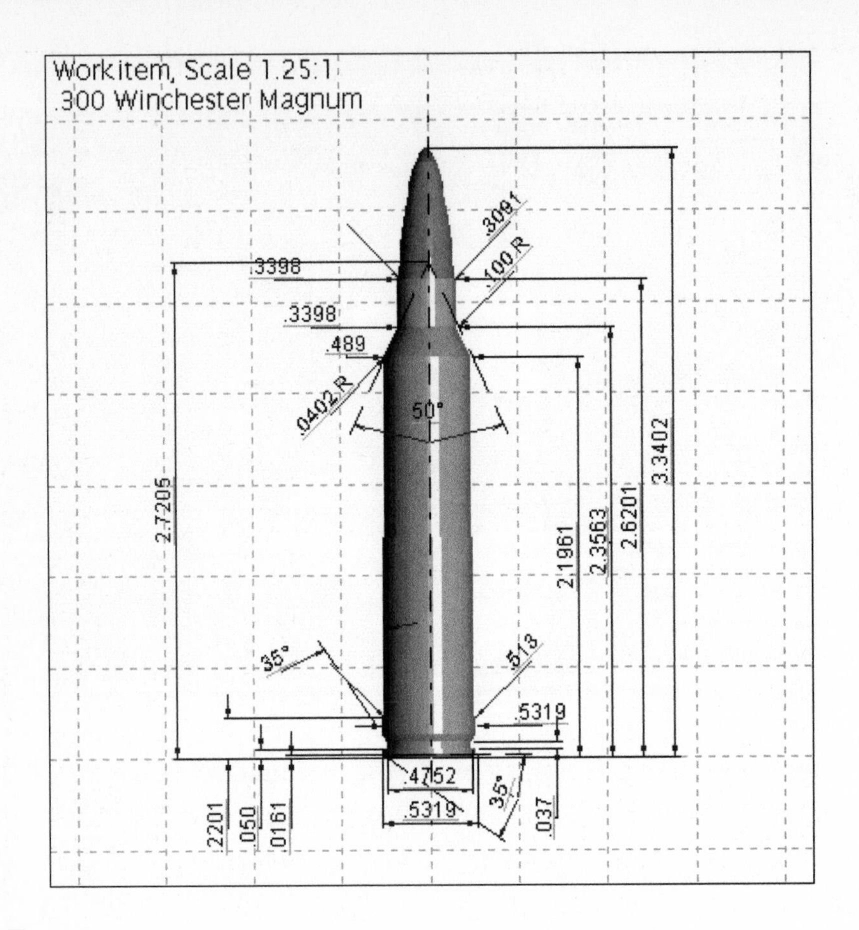


the .338 Lapua is basically a necked down .416 Rigby case that has been reinforced to stand chamber pressures of 60,900 psi! (To use Boston speak: That's freakin' wicked high.)

The most common bullet weights in .338 Lapua sniper rifles are the 250-grain and 300-grain Sierra Match King HPBTs. Lapua also produces a 250-grain Scenar bullet that has a better BC than the Sierra Match Kings. The .338 is said to be effective out to 1,500–1,800 yards, although most shooters would stick with the 1,500–1,600 yard effective range figures. If you feel like geeking out, check out the ballistic tables provided on the next page. A quick look at these tables shows that the 250-grain Match King has a lower BC than the 250-grain Lapua. The Match King goes subsonic between 1,450 and 1,500 yards. The 250-grain Scenar with a BC of 0.675 stays supersonic to 1,750 yards! Again, the bigger BC wins! The 300 -grain Match King, with a BC of 0.768, despite its heavier weight and slower muzzle velocity, retains a supersonic velocity to 1,800 plus yards! You can't escape ballistic science.

Exbal Ballistic Calculator

Federal Gold Medal: 308 Win. : Sierra MatchKing BTHP

| | SIGHT-IN DATA | FIELD DATA |
|---|---|---|
| Muzzle Velocity (fps) | 2600 | 2600 |
| Bullet Weight (grains) | 168 | |
| Sight Height (in) | 1.5 | |
| Sight-in Distance (yd) | 100 | |
| Altitude (ft) | 0 | 0 |
| Temperature (deg F) | 59 | 59 |
| Pressure@Sea Level(in Hg) | 29.53 | 29.53 |
| Relative Humidity (pct) | 78.0 | 78.0 |
| Wind Velocity (mph) | | 0 |
| Wind Angle (degrees) | | 180 ( 6.0 O'Clock) |
| Incline Angle (degrees) | | 0 |
| Moving Target Speed (mph) | | 0.0 |
| Ballistic Coefficient | 0.458 | |

POINT BLANK RANGE DATA

| TARGET HEIGHT (in) | SIGHT-IN DISTANCE (yd) | POINT BLANK RANGE HIGH (yd) | POINT BLANK RANGE LOW (yd) |
|---|---|---|---|
| 2 | 152 | 92 | 171 |
| 4 | 191 | 110 | 220 |
| 6 | 221 | 123 | 258 |
| 8 | 247 | 135 | 289 |
| 10 | 269 | 146 | 315 |

MAX APPARENT TRAJECTORY (in) 0.2

| TARGET DIST (yd) | SIGHT ADJUSTMENTS ELEV MOA | SIGHT ADJUSTMENTS WIND MOA | SIGHT ADJUSTMENTS LEAD MOA | TRAJECTORY VALUES ELEV (in) | TRAJECTORY VALUES WIND (in) | TRAJECTORY VALUES LEAD (in) | VELOCITY (fps) | ENERGY (ft-lb) | Drop from bore line (in) | ARRIVAL TIME (sec) |
|---|---|---|---|---|---|---|---|---|---|---|
| 0 | 0.00 | 0.00 | 0.00 | -1.5 | 0.0 | 0.0 | 2600 | 2521 | 0.0 | 0.0000 |
| 50 | 0.00 | 0.00 | 0.00 | -0.1 | 0.0 | 0.0 | 2504 | 2339 | -0.6 | 0.0588 |
| 100 | 0.00 | 0.00 | 0.00 | -0.0 | 0.0 | 0.0 | 2410 | 2166 | -2.7 | 0.1198 |
| 150 | 1.00 | 0.00 | 0.00 | -1.4 | 0.0 | 0.0 | 2318 | 2004 | -6.2 | 0.1832 |
| 200 | 2.25 | 0.00 | 0.00 | -4.5 | 0.0 | 0.0 | 2228 | 1852 | -11.3 | 0.2492 |
| 250 | 3.50 | 0.00 | 0.00 | -9.3 | 0.0 | 0.0 | 2140 | 1708 | -18.2 | 0.3179 |
| 300 | 5.00 | 0.00 | 0.00 | -16.0 | 0.0 | 0.0 | 2055 | 1574 | -27.0 | 0.3894 |
| 350 | 6.75 | 0.00 | 0.00 | -24.8 | 0.0 | 0.0 | 1971 | 1449 | -37.9 | 0.4639 |
| 400 | 8.50 | 0.00 | 0.00 | -35.7 | 0.0 | 0.0 | 1890 | 1332 | -51.0 | 0.5416 |
| 450 | 10.50 | 0.00 | 0.00 | -49.2 | 0.0 | 0.0 | 1810 | 1222 | -66.5 | 0.6227 |
| 500 | 12.50 | 0.00 | 0.00 | -65.3 | 0.0 | 0.0 | 1733 | 1120 | -84.6 | 0.7073 |
| 550 | 14.50 | 0.00 | 0.00 | -84.2 | 0.0 | 0.0 | 1658 | 1025 | -105.7 | 0.7958 |
| 600 | 17.00 | 0.00 | 0.00 | -106.3 | 0.0 | 0.0 | 1586 | 938 | -129.9 | 0.8882 |
| 650 | 19.50 | 0.00 | 0.00 | -131.9 | 0.0 | 0.0 | 1518 | 859 | -157.6 | 0.9849 |
| 700 | 22.00 | 0.00 | 0.00 | -161.3 | 0.0 | 0.0 | 1452 | 787 | -189.0 | 1.0859 |
| 750 | 24.75 | 0.00 | 0.00 | -194.8 | 0.0 | 0.0 | 1390 | 720 | -224.6 | 1.1915 |
| 800 | 27.75 | 0.00 | 0.00 | -232.7 | 0.0 | 0.0 | 1331 | 661 | -264.6 | 1.3017 |
| 850 | 31.00 | 0.00 | 0.00 | -275.6 | 0.0 | 0.0 | 1278 | 609 | -309.6 | 1.4168 |
| 900 | 34.25 | 0.00 | 0.00 | -323.8 | 0.0 | 0.0 | 1228 | 562 | -359.9 | 1.5366 |
| 950 | 38.00 | 0.00 | 0.00 | -377.7 | 0.0 | 0.0 | 1182 | 521 | -415.9 | 1.6611 |
| 1000 | 41.75 | 0.00 | 0.00 | -437.9 | 0.0 | 0.0 | 1142 | 486 | -478.2 | 1.7903 |
| 1050 | 46.00 | 0.00 | 0.00 | -504.8 | 0.0 | 0.0 | 1106 | 456 | -547.1 | 1.9239 |
| 1100 | 50.25 | 0.00 | 0.00 | -578.7 | 0.0 | 0.0 | 1075 | 431 | -623.1 | 2.0615 |
| 1150 | 54.75 | 0.00 | 0.00 | -660.2 | 0.0 | 0.0 | 1046 | 408 | -706.7 | 2.2032 |
| 1200 | 59.75 | 0.00 | 0.00 | -749.6 | 0.0 | 0.0 | 1020 | 388 | -798.2 | 2.3485 |
| 1250 | 64.75 | 0.00 | 0.00 | -847.4 | 0.0 | 0.0 | 997 | 371 | -898.1 | 2.4974 |
| 1300 | 70.00 | 0.00 | 0.00 | -954.0 | 0.0 | 0.0 | 975 | 355 | -1006.7 | 2.6498 |
| 1350 | 75.75 | 0.00 | 0.00 | -1069.7 | 0.0 | 0.0 | 956 | 341 | -1124.6 | 2.8055 |
| 1400 | 81.50 | 0.00 | 0.00 | -1194.9 | 0.0 | 0.0 | 937 | 328 | -1251.9 | 2.9643 |
| 1450 | 87.50 | 0.00 | 0.00 | -1330.1 | 0.0 | 0.0 | 920 | 316 | -1389.2 | 3.1262 |
| 1500 | 94.00 | 0.00 | 0.00 | -1475.6 | 0.0 | 0.0 | 905 | 305 | -1536.8 | 3.2911 |
| 1550 | 100.50 | 0.00 | 0.00 | -1631.8 | 0.0 | 0.0 | 890 | 295 | -1695.1 | 3.4589 |
| 1600 | 107.25 | 0.00 | 0.00 | -1799.1 | 0.0 | 0.0 | 876 | 286 | -1864.4 | 3.6296 |
| 1650 | 114.50 | 0.00 | 0.00 | -1977.8 | 0.0 | 0.0 | 862 | 277 | -2045.2 | 3.8031 |
| 1700 | 121.75 | 0.00 | 0.00 | -2168.3 | 0.0 | 0.0 | 849 | 269 | -2237.8 | 3.9794 |
| 1750 | 129.50 | 0.00 | 0.00 | -2371.0 | 0.0 | 0.0 | 836 | 260 | -2442.6 | 4.1587 |
| 1800 | 137.25 | 0.00 | 0.00 | -2586.3 | 0.0 | 0.0 | 823 | 253 | -2659.9 | 4.3408 |
| 1850 | 145.25 | 0.00 | 0.00 | -2814.6 | 0.0 | 0.0 | 811 | 245 | -2890.4 | 4.5258 |
| 1900 | 153.50 | 0.00 | 0.00 | -3056.3 | 0.0 | 0.0 | 799 | 238 | -3134.2 | 4.7138 |
| 1950 | 162.25 | 0.00 | 0.00 | -3311.9 | 0.0 | 0.0 | 788 | 232 | -3391.9 | 4.9046 |
| 2000 | 171.00 | 0.00 | 0.00 | -3581.8 | 0.0 | 0.0 | 777 | 225 | -3663.8 | 5.0984 |

December 05, 2009

**308 Win Ballistic calculator**

The .338 Lapua is what most people would consider an intermediate sniper cartridge—heavier and more powerful than the .308 and .300 Win Mag but not in the .408 CheyTac and 5.0 BMG league. The CheyTac has a muzzle energy of 7,700 pound-feet and the 50 BMG an amazing muzzle energy of 12,000 pound-feet, while the .338 Lapua has 5,222 pound-feet for the 300 grain Match King. So the .338 can be used on some hard targets but will not have the penetrating power of the .408 or the .50.

The recoil is very manageable if the rifle is fitted with a muzzle brake. The .338 Lapua from Accuracy International is a joy to shoot and very accurate, with sub .5 MOA at 100 yards. Shooting 4-inch clay birds at 1,000 yards is no problem with this rifle.

Exbal Ballistic Calculator

Black-Hills New Mfg: .223 Remington : Sierra MatchKing 0.224" 77gr HPBT

| | SIGHT-IN DATA | FIELD DATA | | |
|---|---|---|---|---|
| Muzzle Velocity (fps) | 2750 | 2750 | | |
| Bullet Weight (grains) | 77 | | | |
| Sight Height (in) | 1.5 | | | |
| Sight-in Distance (yd) | 100 | | | |
| Altitude (ft) | 0 | 0 | | |
| Temperature (deg F) | 59 | 59 | | |
| Pressure@Sea Level(in Hg) | 29.53 | 29.53 | | |
| Relative Humidity (pct) | 78.0 | 78.0 | | |
| Wind Velocity (mph) | | 0 | | |
| Wind Angle (degrees) | | 180 ( 6.0 O'Clock) | | |
| Incline Angle (degrees) | | 0 | | |
| Moving Target Speed (mph) | | 0.0 | | |
| Ballistic Coefficient | 0.372 | 0.362 | 0.362 | 0.343 |
| Lower Velocity Limit (fps) | 3000 | 2500 | 1700 | 0 |

POINT BLANK RANGE DATA

| TARGET HEIGHT (in) | SIGHT-IN DISTANCE (yd) | POINT BLANK RANGE HIGH (yd) | POINT BLANK RANGE LOW (yd) |
|---|---|---|---|
| 2 | 158 | 96 | 178 |
| 4 | 198 | 114 | 227 |
| 6 | 228 | 127 | 264 |
| 8 | 254 | 139 | 297 |
| 10 | 276 | 153 | 321 |

MAX APPARENT TRAJECTORY (in) 0.1

| TARGET DIST (yd) | SIGHT ADJUSTMENTS ELEV MOA | WIND MOA | LEAD MOA | TRAJECTORY VALUES ELEV (in) | WIND (in) | LEAD (in) | VELOCITY (fps) | ENERGY (ft-lb) | Drop from bore line (in) | ARRIVAL TIME (sec) |
|---|---|---|---|---|---|---|---|---|---|---|
| 0 | 0.00 | 0.00 | 0.00 | -1.5 | 0.0 | 0.0 | 2750 | 1293 | 0.0 | 0.0000 |
| 50 | 0.25 | 0.00 | 0.00 | -0.1 | 0.0 | 0.0 | 2625 | 1178 | -0.6 | 0.0558 |
| 100 | 0.00 | 0.00 | 0.00 | -0.0 | 0.0 | 0.0 | 2503 | 1071 | -2.4 | 0.1143 |
| 150 | 0.75 | 0.00 | 0.00 | -1.3 | 0.0 | 0.0 | 2384 | 972 | -5.6 | 0.1757 |
| 200 | 2.00 | 0.00 | 0.00 | -4.1 | 0.0 | 0.0 | 2269 | 880 | -10.4 | 0.2401 |
| 250 | 3.25 | 0.00 | 0.00 | -8.5 | 0.0 | 0.0 | 2157 | 795 | -16.8 | 0.3079 |
| 300 | 4.75 | 0.00 | 0.00 | -14.9 | 0.0 | 0.0 | 2049 | 717 | -25.2 | 0.3792 |
| 350 | 6.25 | 0.00 | 0.00 | -23.3 | 0.0 | 0.0 | 1943 | 646 | -35.5 | 0.4544 |
| 400 | 8.00 | 0.00 | 0.00 | -34.0 | 0.0 | 0.0 | 1842 | 580 | -48.2 | 0.5336 |
| 450 | 10.00 | 0.00 | 0.00 | -47.3 | 0.0 | 0.0 | 1743 | 519 | -63.5 | 0.6173 |
| 500 | 12.00 | 0.00 | 0.00 | -63.5 | 0.0 | 0.0 | 1645 | 463 | -81.6 | 0.7058 |
| 550 | 14.50 | 0.00 | 0.00 | -82.8 | 0.0 | 0.0 | 1551 | 411 | -102.9 | 0.7997 |
| 600 | 16.75 | 0.00 | 0.00 | -105.8 | 0.0 | 0.0 | 1462 | 366 | -127.8 | 0.8992 |
| 650 | 19.50 | 0.00 | 0.00 | -132.8 | 0.0 | 0.0 | 1379 | 325 | -156.8 | 1.0048 |
| 700 | 22.50 | 0.00 | 0.00 | -164.4 | 0.0 | 0.0 | 1303 | 290 | -190.3 | 1.1167 |
| 750 | 25.50 | 0.00 | 0.00 | -201.1 | 0.0 | 0.0 | 1235 | 261 | -229.0 | 1.2350 |
| 800 | 29.00 | 0.00 | 0.00 | -243.5 | 0.0 | 0.0 | 1174 | 236 | -273.4 | 1.3596 |
| 850 | 32.75 | 0.00 | 0.00 | -292.2 | 0.0 | 0.0 | 1122 | 215 | -324.0 | 1.4903 |
| 900 | 37.00 | 0.00 | 0.00 | -347.8 | 0.0 | 0.0 | 1079 | 199 | -381.6 | 1.6267 |
| 950 | 41.25 | 0.00 | 0.00 | -410.9 | 0.0 | 0.0 | 1041 | 185 | -446.6 | 1.7683 |
| 1000 | 46.00 | 0.00 | 0.00 | -481.9 | 0.0 | 0.0 | 1007 | 173 | -519.7 | 1.9150 |
| 1050 | 51.00 | 0.00 | 0.00 | -561.6 | 0.0 | 0.0 | 978 | 163 | -601.3 | 2.0662 |
| 1100 | 56.50 | 0.00 | 0.00 | -650.4 | 0.0 | 0.0 | 951 | 155 | -692.0 | 2.2219 |
| 1150 | 62.25 | 0.00 | 0.00 | -748.7 | 0.0 | 0.0 | 927 | 147 | -792.3 | 2.3818 |
| 1200 | 68.25 | 0.00 | 0.00 | -857.2 | 0.0 | 0.0 | 906 | 140 | -902.8 | 2.5458 |
| 1250 | 74.50 | 0.00 | 0.00 | -976.4 | 0.0 | 0.0 | 886 | 134 | -1023.9 | 2.7136 |
| 1300 | 81.25 | 0.00 | 0.00 | -1106.6 | 0.0 | 0.0 | 867 | 128 | -1156.1 | 2.8852 |
| 1350 | 88.25 | 0.00 | 0.00 | -1248.5 | 0.0 | 0.0 | 849 | 123 | -1299.9 | 3.0606 |
| 1400 | 95.75 | 0.00 | 0.00 | -1402.5 | 0.0 | 0.0 | 831 | 118 | -1455.9 | 3.2398 |
| 1450 | 103.25 | 0.00 | 0.00 | -1569.1 | 0.0 | 0.0 | 815 | 113 | -1624.5 | 3.4229 |
| 1500 | 111.25 | 0.00 | 0.00 | -1749.0 | 0.0 | 0.0 | 798 | 109 | -1806.4 | 3.6098 |
| 1550 | 119.75 | 0.00 | 0.00 | -1942.7 | 0.0 | 0.0 | 783 | 105 | -2002.0 | 3.8005 |
| 1600 | 128.25 | 0.00 | 0.00 | -2150.7 | 0.0 | 0.0 | 768 | 101 | -2211.9 | 3.9952 |
| 1650 | 137.25 | 0.00 | 0.00 | -2373.6 | 0.0 | 0.0 | 754 | 97 | -2436.8 | 4.1937 |
| 1700 | 146.75 | 0.00 | 0.00 | -2612.0 | 0.0 | 0.0 | 741 | 94 | -2677.2 | 4.3961 |
| 1750 | 156.50 | 0.00 | 0.00 | -2866.6 | 0.0 | 0.0 | 728 | 90 | -2933.7 | 4.6025 |
| 1800 | 166.50 | 0.00 | 0.00 | -3137.9 | 0.0 | 0.0 | 715 | 87 | -3207.0 | 4.8128 |
| 1850 | 176.75 | 0.00 | 0.00 | -3426.6 | 0.0 | 0.0 | 703 | 84 | -3497.7 | 5.0271 |
| 1900 | 187.75 | 0.00 | 0.00 | -3733.4 | 0.0 | 0.0 | 691 | 82 | -3806.4 | 5.2453 |
| 1950 | 198.75 | 0.00 | 0.00 | -4058.9 | 0.0 | 0.0 | 680 | 79 | -4133.9 | 5.4677 |
| 2000 | 210.25 | 0.00 | 0.00 | -4403.9 | 0.0 | 0.0 | 668 | 76 | -4480.8 | 5.6943 |

December 05, 2009

## The CheyTac .408

If you are impressed with the potential capabilities of the .338, you better sit down, strap in, and hold on for the .408. The story of CheyTac began a few years ago. John D. Taylor saw that there was a need to either improve the performance of the .50 BMG, which had been developed

**Brandon in the mountains of Northern Iraq. Good CheyTac country.**

80 to 90 years ago. (BMG is short for Browning machine gun. The cartridge was originally designed for use in a machine gun—not as a sniper cartridge.) Taylor brought together a team and created CheyTac Associates. Their mission was to develop and build a very long-range sniper cartridge. The CheyTac team also saw a need for a cartridge that would fill the gap between the .338 Lapua and the .50 BMG, especially for soft target (human) engagements.

The CheyTac team approached their mission in an unconventional manner—first they designed the ideal bullet, then designed the rifle to shoot the bullet. They realized that their bullet would have to possess outstanding ballistics at distance. CheyTac developed a concept that is known as Balanced Flight, which they patented. The bullets were developed by Warren Jensen, a partner and designer, at Lost River Ballistic Technologies and were designed using PRODAS software and the concepts of Balanced Flight.

Jensen and associates designed a bullet where the linear drag is matched to the rotational drag, which is one of the concepts of Balanced Flight. It is a well-known ballistic fact that a bullet that can retain its stability throughout its flight will go farther and be more accurate. Additionally, CheyTac designers determined the ideal design of the barrel's lands and grooves. They determined that the ratio of total surface of the bullet to the total surface of the lands and grooves should be somewhere between 3 and 4 to 1. Jensen experimented with a variety of calibers but found improvements in bullet performance for the .408 were far beyond predicted.

CheyTac designers also looked into the bullet composition. All of the bullets that we have talked about so far are composed of a lead core with a full copper jacket. The CheyTac bullets are made of a copper nickel alloy and are actually made on a CNC machine (computer controlled) and lathe turned. These bullets are often referred to as solids.

As you can see, the 305-grain bullet has a muzzle velocity of 3,300 fps at the muzzle. The term "max ord" is short for minimal maximal ordinate. When a bullet leaves the muzzle of the rifle, gravity immediately begins to exert its force on the bullet. It is interesting to note that if you were to have your rifle perfectly level on a bench at your local range then simultaneously fire a round and drop a cartridge from the bench, the bullet leaving your gun and the cartridge dropped from the bench

# THE LONGEST CONFIRMED KILL WITH A 7.62 RIFLE IN IRAQ

## U.S. Army Sniper Reportedly Nails A Record 1250 Meter Shot With A 7.62mm Remington M24 SWS

◂ **Remington M24 SWS (Sniper Weapon System)**
Since its adoption by the US Army, the M24 is the standard by which all other military grade sniper rifles are judged. The M24 is world renowned as a long-range precision sniper system capable of enduring the harshest of military environments to include extreme high altitude and the depths of the ocean. The M24 is a combat proven force multiplier serving the US Army with honor and distinction against our nation's enemies; past, present and future.

would hit the ground at the same time. Skeptical? The next time you are at the range, pay attention to the angle of your rifle. It may look level, but the scope is actually dialing in elevation and thereby changing the path of the bullet from a straight line to an arc or trajectory. As you dial in elevation your rifle is gradually angling up to compensate for the increased range to your targets. At long ranges it's almost like you're "lobbing" your bullets into the target. The maximum height of the bullet's trajectory above the bore is known as the maximum ordinate. Is knowing your bullet's maximum ordinate important? Either "yes" or "no" is correct, depending on the context. If you are always shooting in wide open country, with nothing between you and your target except empty sky, then knowing your max ordinate is not important. However, if you are shooting at a target that is below a series of overhead structures, such as a bridge or tree cover, knowing your max ordinate *is* important. You may think you have a clear shot, but if what's overhead happens to be at the right distance and is the right height when your bullet's trajectory reaches its max ordinate, the bullet will impact the structure—not your target. You may not have seen the offending object in your scope depending on the magnification, but it was there. Your carefully planned shot was missing a tiny detail—the max ordinate of the bullet—causing you to kill the bridge or someone on the bridge, while your real target flips you off, runs for cover, or tries to return fire. Not a good thing.

The CheyTac 305 grain bullet is designed to have a max ordinate of only a few feet over the 1,000-yard

**The McMillan TAC-300 is your friend when you have a long way to travel.**

Photo Credit: McMillan

*Photo Credit: Lewis Page,* The Register

**British Snipers have easy access to Accuracy International's excellent rifles. Shown here is the L115A3 .338 arctic warfare model. All rifles are made in Portsmouth, England.**

distance, so a sniper could theoretically hold center mass on a human target from one to 1,000 yards and effectively engage his target. Not too shabby.

## The CheyTac Intervention Rifle

The M-200 version of the CheyTac Intervention Rifle is a bolt-action rifle with a 7-shot capacity detachable magazine. The rifle and the 419-grain bullet are proven out to 2,000 plus yards on soft target, antipersonnel missions. CheyTac claims sub 1 MOA accuracy out to 3,000 yards in testing. Groups of 7 to 9 inches at 1,000 yards, 10 inches at 1,500 yards, and 15 inches at 2,000 yards have been consistently obtained. Personal experience has seen the rifle shoot a 5-inch group at 1,000 yards and have first round hits on a 12-by-20-inch plate at 2,200 yards.

This extreme long-distance capability gives the sniper great standoff distance compared to other cartridges. In testing in Idaho, an observer at the target could not see the shooter in the open on the desert floor at 2,000 yards. Adding a suppressor, the sniper—without any camouflage—

could not be seen with binoculars at this distance.

The rifle repeats its accuracy and holds its zero very well. The system can be disassembled and reassembled with no change in the zero. This includes removal of the barrel, removing the optics and the suppressor, then putting everything back together and shooting again. We have seen this capability with the CheyTac and the Accuracy International systems.

The CheyTac bullet has excellent penetration properties. It can penetrate Level IIIA armor at 2,000 yards and a cinder block wall at 500 plus yards. It will penetrate 1-inch cold-rolled steel at 200 yards and 0.5-inch cold-rolled steel at 850 yards. The 408 system as an antipersonnel weapon is limited only by flight time to its intended target. In testing at the Yuma Testing Grounds, potentially lethal engagements of simulated soft targets (gelatin) were consistently made even at subsonic velocities because of the excellent stability of the projectile through the transition into subsonic velocity.

Accuracy International's Super Magnum .338 Lapua

As an antimaterial weapon it's also very effective. The .50 BMG has an initial higher muzzle energy of about 11,200 to 12,000 pound-feet compared to the .408's muzzle energy of 7,700 pound-feet. However, at about 700 yards, the remaining energy of the .408 is higher than the .50. CheyTac claims that the 419-grain 408 projectile will defeat any material that the .50 BMG can except for those targets that require an explosive projectile, such as the Raufoss round. Materials such as jet engines, engine blocks, or surface missiles can be easily engaged and defeated with solid projectiles such as the CheyTac copper/nickel alloy bullet.

## The .50-Caliber BMG

The .50 BMG was originally developed in World War I as an antimaterial round. Shortly after the Korean War, individuals began experimenting with developing a shoulder-fired rifle that could safely handle the awesome power of this round, which was initially meant for weapons mounted on a vehicle or used on a heavy tripod. The early results weren't very good, but over time the .50 BMG caliber ammunition and rifle design

Photo Credit: CheyTac

# CheyTac® M-200

The Reference Standard Bolt Action Rifle in the 408 CheyTac® Caliber

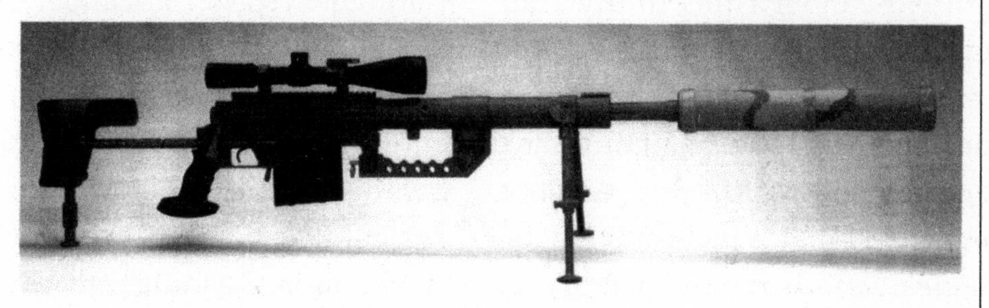

**Features:**

- CNC Machined, Receiver
- Attachable Picatinny Rail M-1913
- Detachable Barrel
- Integral Bipod
- 3.5 lb. Trigger Pull
- Highly Effective Muzzle Brake

**Specifications:**

| | |
|---|---|
| Action Type | • Ultra Heavy Bolt |
| Caliber | • 408 CheyTac® |
| Sights | • None, Scope Rail Provided |
| Overall Length | • 55 inches (stock extended) |
| Barrel Length | • 29 inches |
| Magazine Cap. | • 7 Rounds |
| Weight | • 27 lbs. |
| Stock | • Retractable |

CheyTac M-200

evolved to sniper-level accuracy. Civilian competition shooters with hand-loaded ammo have shot .25 to .5 MOA groups at 1,000 yards—that is a 2.5- to 5.0-inch group at 1,000 yards! Like the CheyTac bullet, the bullets in these world record shots are solid bullets that are turned on a lathe. The downside of these solid projectiles is faster barrel wear and, of course, cost.

Standard military .50 BMG ammo is not that accurate. This is not surprising given the fact that the bulk of military .50-caliber ammo is manufactured to be used in a machine gun, which is designed to shoot a pattern, not a sub 1 MOA group. Newer, more accurate military ammunition is available and indeed, there have been several extreme long-distance confirmed kills made with .50-caliber rifles in the Iraq and Afghanistan theaters. One of the newer, more accurate rounds is known as the saboted light armor penetrator (SLAP) round. A sabot (or "shoe") is a sleeve, usually plastic in rifles, that surrounds a bullet that is smaller than the actual diameter of the barrel, enabling the round to exit the barrel at much higher than normal velocities. After the round leaves the barrel the sabot falls away, and the smaller, lighter round rockets toward its target. The reported muzzle velocity is close to 4,000 fps and has the capability of penetrating .75-inch steel at 1,500 yards. Shooting with that type of muzzle velocity produces a nice flat trajectory, with a low max ord, and that type of penetrating power at distance makes it extremely dangerous.

Another type of .50-caliber round which has been called the crown jewel of .50 caliber is the Raufoss Multipurpose round. The Raufoss was developed in Norway and its design incorporates an exploding, armor-piercing tungsten carbide penetrator. During the bullet's acceleration to the intended target the incendiary compound compresses, which makes a small air pocket. Upon impact the air pocket is compressed, leading to the ignition of the explosive mixture. This then sets off a tiny explosive charge. As the tungsten penetrator jolts and fragments out of the core of the bullet, white-hot sparks of zirconium particles follow and are capable of

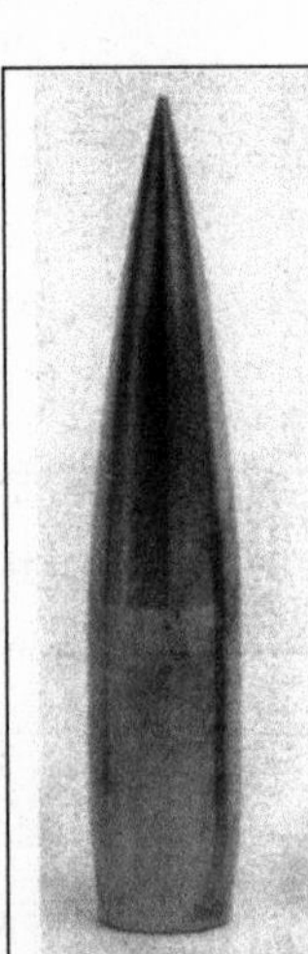

**.408 CheyTac Caliber**
**419 grain bullet:**

Magnum rifle primer
Bullet diameter: .4077"
Copper nickel alloy
Overall cartridge length: 4.307
Velocity: 2900 fps
198 rounds per ammo can
Ballistic Coefficient: .94 (avg. over 3500 yards)

*Photo Credit: CheyTac*

The CheyTac .408 standard bullet

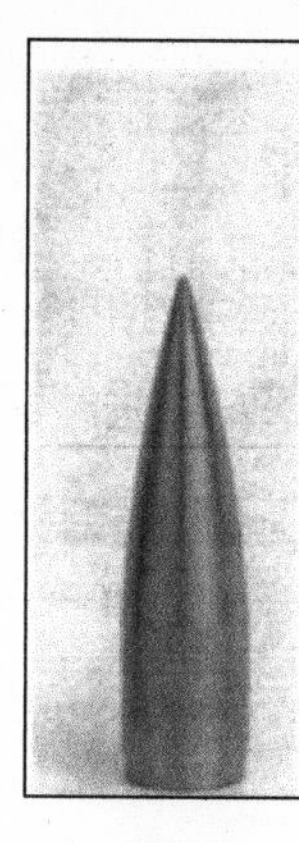

**.408 CheyTac Caliber**
**305 grain bullet:**

Magnum rifle primer
Diameter: .4077
Copper nickel alloy
Overall length: 4.100"
Velocity: 3300 fps
198 rds. per can
BC is not applicable since this round is only meant to go 1000 yds. and closer at a high rate of speed and MINIMUM max. ord.

*Photo Credit: CheyTac*

For closer ranges, they developed the faster-and flatter-shooting 305 grain.

igniting fuel or explosive vapors in the area of bullet contact. According to its developer, the end result is "the Nordic Ammunition Company, (NAMMO), equivalent firing power of a .20mm projectile" and due to its penetration and delayed detonation it "moves projectile fragmentation and damage effect inside the target for maximum anti-personnel and fire start effect." The Raufoss bullet also happens to be very accurate.

In Dean Michaelis's book *The Complete .50 Caliber Sniper Course*, he notes that even if you are a great shot with a lower-caliber sniper rifle, technique becomes paramount as you step up to the big dogs. The .50-caliber rifles must be shot neutral. That means that you need to have your natural point of aim, be straight behind the gun, and not try to muscle the gun to your intended target. Trying to force the gun will result in a miss. Finesse and technique will bring you home with a hit.

The U.S. military has the following .50-caliber cartridges in its inventory:

- M-33 Ball—not very accurate.
- M-17 Tracer.
- M-8 Armor Piercing Incendiary. It is reportedly reasonably accurate.
- M-20 Armor Piercing Incendiary—a tracer shot for the M-8.
- M-903 Saboted Light Armor Piercing. Composed of a 350-grain tungsten steel penetrator. This requires a different twist rate than other standard rounds.
- M-933 SLAP Tracer—a tracer for the M-903.
- Mark 211 , Mod O. This round was developed for and adapted by the U.S. Navy Special Warfare snipers. It is also referred to as a greentip.

The selection of different cartridges will depend upon the mission requirements, and each round will

Photo Credit: CheyTac

The CheyTac Intervention Rifle

have a different ballistic profile, which the shooter will need to know. In the longer distances (1,000 plus yards) that the .50 caliber can be employed, for analysis and accurate interpretation of the meteorological and environmental factors become very important in making accurate hits. The use of a ballistic and the input of environmentals and accurate ballistic data can ensure a first-round hit.

Photo Credit: CheyTac

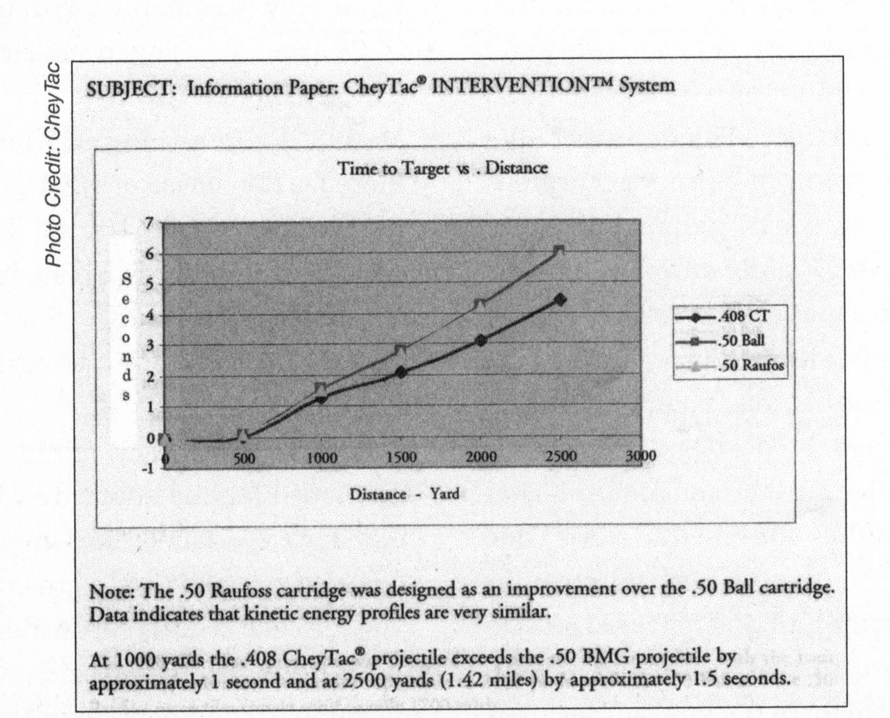

SUBJECT: Information Paper: CheyTac® INTERVENTION™ System



Note: The .50 Raufoss cartridge was designed as an improvement over the .50 Ball cartridge. Data indicates that kinetic energy profiles are very similar.

At 1000 yards the .408 CheyTac® projectile exceeds the .50 BMG projectile by approximately 1 second and at 2500 yards (1.42 miles) by approximately 1.5 seconds.

CheyTac information paper

Look at the penetration from the chart below and understand that this is just the standard AP round, not the SLAP or Raufoss. Now you can understand why the .50 may not be the best choice to use when concerned about overpenetration. We don't want any friendlies in the house three blocks away from the target to get hit as well. But for the 2,000 plus yard shot in the rocky mountains of Afghanistan, this dog will hunt.

## What Can Mess You Up: Accuracy Issues

Anyone passionate about shooting wants to do it well, and doing it well means *accuracy*, and maybe throw *speed* in there too. So many things can influence the shot placement; however, not all may be in the forefront of your mind. Here are some things other than environmentals and ballistics to think about:

- Hopefully, you have been keeping a record of the number of rounds you have put through your rifle. Over time, barrels will lose their accuracy. This is secondary to the degradation of the rifling and the throat of the barrel. The throat is the area just forward of the bullet when the bullet is in the chamber. This is the part of the rifle barrel that takes

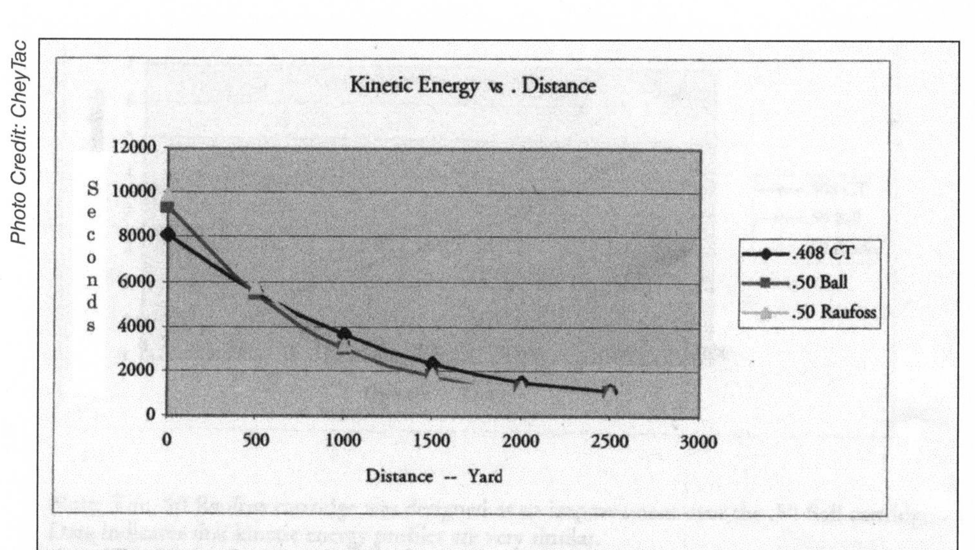

Photo Credit: CheyTac

Note: The .50 Raufoss cartridge was designed as an improvement over the .50 Ball cartridge. Data indicates that kinetic energy profiles are very similar.

The .408 CheyTac® kinetic energy surpasses the .50 Ball and the .50 Raufoss at approximately 400 yards and maintains the lead beyond 2500 yards (1.42 miles).

In supervised tests, the .408 CheyTac® projectile penetrated armor and laminated glass that was resistant to the .50 BMG projectiles (US Armed Forces Journal, August 2003)

the brunt of the high pressure and heat of the round being fired. Some rifle cartridges will wear faster than others. In general, the hotter loads will have a shorter barrel life than the slower rounds.

- Your scope can become loose. The repeated stress of the rifle firing can loosen the scope mount, which will be just enough to have the scope move slightly on subsequent shots. Unless caught, that will ruin your day.
- The crown of your rifle barrel can be damaged. The crown is the area just at the very end of the rifle at the muzzle. It is the last metal part of the barrel the bullet touches before it leaves the rifle. As the bullet leaves the barrel, all of the hot gases that have been pushing it down the barrel escape. If these gases do not escape evenly around the bullet's base, the resulting uneven pressures can influence the flight of the bullet and contribute to the bullet's stability, which affects the accuracy of the bullet. If the crown gets chipped or worn your rifle's accuracy can suffer. Usually you see flyers in your groups.
- Parallax problems. To see parallax, set your rifle up on sandbags so that it is solidly placed. Run your scope up to max power and line up the crosshair perfectly on a bull's eye at one hundred yards. Without touching the rifle, look at the crosshair and move your eye/ head slightly side to side, as well as up and down. If the crosshair

**50-Caliber AP Round Short-Range Media Penetration (in inches)**

| Medium | 200 meters | 600 meters | 1,500 meters |
|---|---|---|---|
| Sand (100 lb dry weight) | 14 | 12 | 6 |
| Clay (100 lb dry weight) | 28 | 26 | 21 |
| Concrete | 2 | 1 | 1 |
| Armor plate (homogeneous) | 1.0 | 0.7 | 0.3 |
| Armor plate (face hardened) | 0.9 | 0.5 | 0.2 |

*Photo Credit: CheyTac*

# CheyTac® / 408

**HIgher Performance Is The Future of Ammunition**

CheyTac® 408/419gr. (M40) vs. Current Issue Military M8 AP
Range 650 Measured Yards
Steel Plate 1/2 Inch Thickness

While the 50 cal. M8 AP ammunition did create an opening in the plate at 650 yds., it is clear that the round did defeat the plate. The penetration is dramatically sub caliber and the front of the plate shows the projectile's energy dissipation. The 408, 419 gr., M40 round shows a full caliber, clean penetration at the same measured distance. The interesting point is the M40, 419 gr. 408 CheyTac® round is not primarily intended as an AP round. It is the standard, long range, soft target interdiction round. CheyTac® produces multiple types of AP rounds for the 408 which have added and unique capabilities for the Military Professional.

The M40 .408, 419 gr. will defeat 90% of the targets that the average sniper is likely to see, whether it be steel, glass, ceramic, or other materials. The only things that the M40 won't defeat are Rolled Homogeneous Armor and 4" armored glass. It should be noted, even 50 cal AP ammo will not defeat these targets. If the operator has to face this kind of threat he will need a true anti-tank munition.

The only types of targets which could present this armor profile is a heavy armored car or truck. 408 or 50 BMG will not be sufficient to stop these targets cold. A LAW or other anti-tank weapon would be needed.

The critical issue in ammunition and weapons facing the dedicated sniper operator today, is performance. At CheyTac®, we are dedicated to the highest level of performance in our design and manufacturing. This exacting level of performance is your advantage in the real world of opeations.

**.50-Caliber AP Round Media Penetration at 100 yards (in inches).**

| Medium | Inches |
|---|---|
| Concrete (solid) | 9 |
| Timber logs | 96 |
| Steel ( non-armored) | 1.8 |
| Aluminum | 3.5 |
| Tamped snow | 77 |
| Dry soil | 28 |
| Wet soil | 42 |
| Dry sand | 24 |
| Wet sand | 36 |
| Dry clay | 42 |
| Wet clay | 64 |

*Photo Credit: Data from John Plasters, The Ultimate Sniper.*

**Advantages of a Sniper-Grade Bolt-Action Rifle:**

- Degree of Accuracy—.5 MOA accuracy.
- Very Reliable—rare malfunction, secondary to design and number of parts in the firing mechanism. The only real parts of the rifle that are going to fail in the short term are the trigger, the firing pin in the bolt, or possibly the bolt could lock up if using low-grade or bunk ammunition.
- Ease of field maintenance.

moves from the bull's eye, you have a parallax problem. This can be corrected by the scope manufacturer, but it must be corrected for a specific distance. Most sniper-quality scopes will have a parallax adjustment knob.

- Fouling of the barrel. Copper fouling is a close second to loose scope mount screws. Copper fouling is hard to see and hard to pinpoint. It occurs slowly and what you will typically see is your groups open up over time. If you

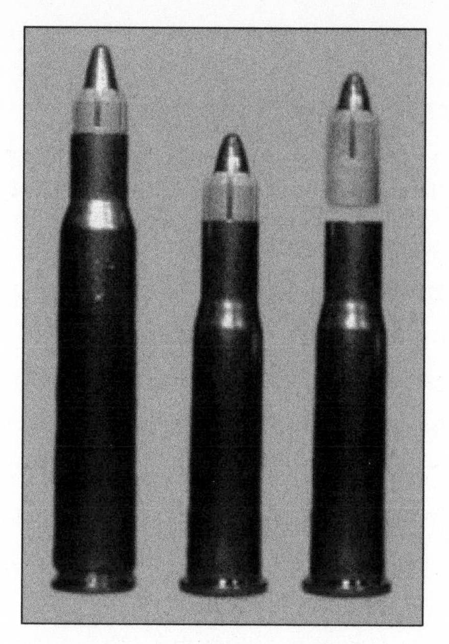

Example of a sabot plastic sleeve

Photo Credit: The Complete .50 Caliber Sniper Course
by Dean Michaels


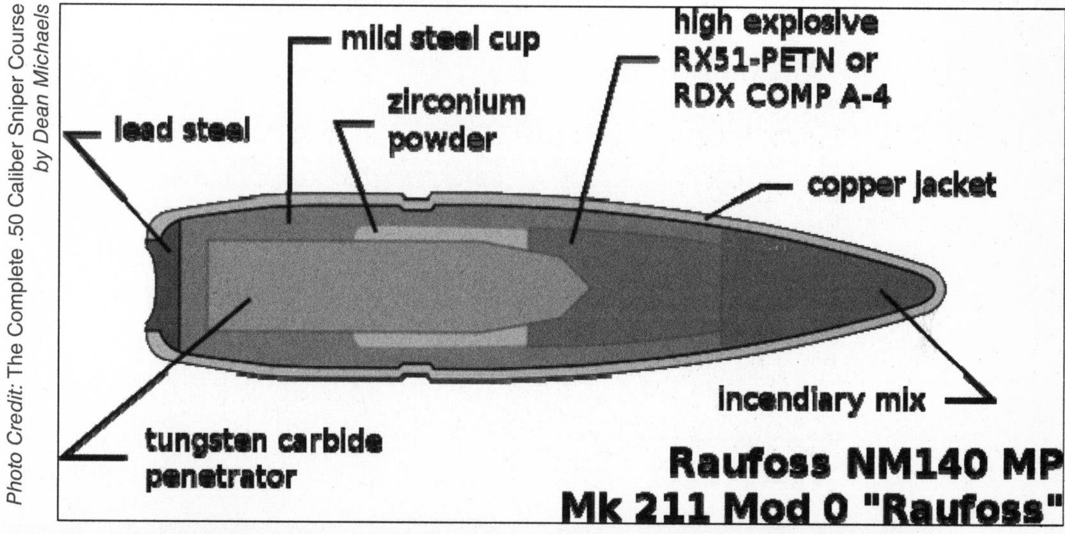

Cut away of an Accuracy International Rifle Receiver and bolt

**Disadvantages of a Bolt-Action Sniper-Grade Rifle:**

- Rate of fire. Depending on the manufacturer, the rifle may have an internal magazine, which in the lighter class rifles limits the shooter to five rounds.
- Some manufacturers, such as CheyTac, Accuracy International, and McMillan, have addressed this issue by designing their rifles to use detachable box magazines that can hold more rounds and be exchanged for a fresh magazine as needed. The CheyTac magazine will hold seven rounds; typical 308 magazines will hold up to ten. Despite this increased ammunition capacity, a bolt rifle will never have the same rate of fire as a semiautomatic. Why? Manually working the bolt takes some time, that's why.

**Advantages of a Sniper-Grade Semiautomatic Rifle:**

- Rate of fire. The semiautomatic action will cycle the bolt much faster than we humans can ever hope to.

**Disadvantages of a Sniper-Grade Semiautomatic Rifle:**

- Degree of accuracy. The best semiautomatics will shoot 1 MOA in the lighter calibers. In the heavy caliber rifles such as the Accuracy International .50 caliber, the accuracy is closer to 1.5 MOA. Granted this is very good, but depending on the mission and the target it may not be good enough.
- Dependability. There are many more moving parts in the semiautomatic rifles. Any of these parts can fail. Additionally, the semiautomatics have a higher chance of a feeding/ejection malfunction than a bolt gun.
- Ease of maintenance. More moving parts, more to clean, and in gas impingement semiautomatics, the carbon build-up is another thing you need to keep a close eye on.

*Photo Credit: Photo by Curtis Prejean, courtesy of Accuracy International*

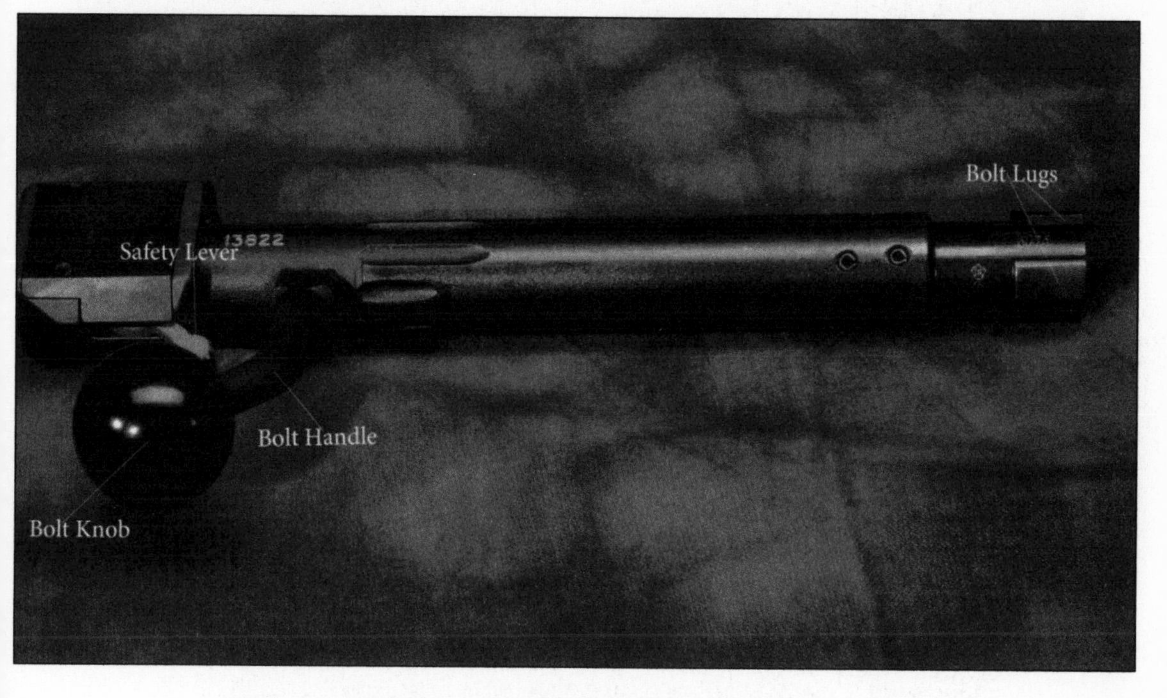


**Check out the AR style rifle and compare it to the simplicity of a bolt action.**

*Photo Credit: Photo by Curtis Prejean, courtesy of Accuracy International*

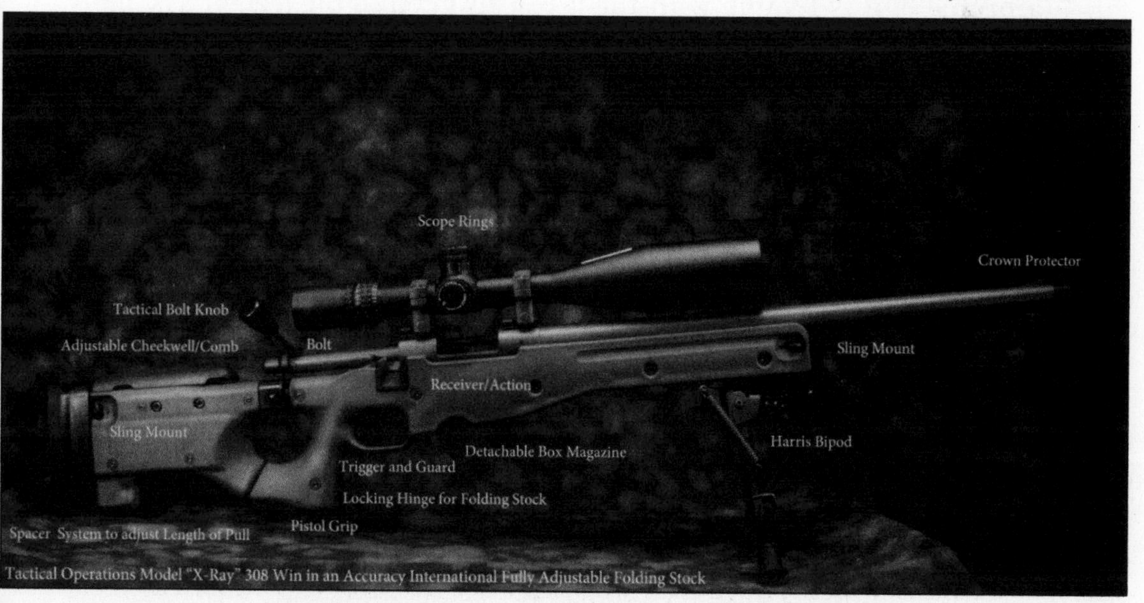


**AR style upper and lower parts schematics. That's a lot of parts.**

Brandon's 300 Winmag with NF optics. With a 2,900 feet per second muzzle velocity this is a real tack driver out to 1,000 yards.

determine that copper fouling is a problem, it is fixable with the available copper removal chemical solutions. Be careful with the aggressive solutions. If left in contact with your barrel for too long (15 minutes or more), they can damage the steel as well as remove the copper. Additionally, be careful not to let these chemicals get into the trigger assembly of your rifle.

- Your vision. Stay on top of it boys and girls: Wear your eye protection religiously, and don't be too proud to stop in and say hello to your local optometrist now and again.

8

# Suppressors

During coauthor Glen Doherty's third phase of Basic Underwater Demolition/SEAL (BUD/S) training, he made several laps in the ocean while training out at San Clemente Island because he couldn't stop himself from calling a magazine a "clip." Too much *T. J. Hooker* as a kid I guess.

The same goes for most of the general public not in the know, who often refer to a suppressor as a silencer and actually believe that a long-barreled rifle fitted with a "silencer" would make no more noise than your one year old passing gas. Not so much. There really is no such thing as a silencer, so from here on out, it's suppressor only, understand? Now, this is just one guy's opinion and is not shared by the U.S. Department of Justice, or the ATF for that matter, who still refer to the mechanism as a silencer. We're trying to start a grass roots movement. Roger? OK. So what gets suppressed? Pretty much everything coming out the end of the barrel. Gas, flash, noise, speed . . . everything.

The modern suppressor dates back to the early 20th century, and its development is credited to Hiram Maxim, who also helped develop mufflers for cars.

Photo Credit: ACC Remington

Advanced Armament Corp billboard. AAC makes fantastic suppressors that can be found on a lot of weapons. Suppressors are legal in 36 states.

Shooters Depot Advanced Rifle Integral External Suppressor (ARIES) has been called the "holy grail of silencers." It uses carbon fiber tubing.

Photo Credit: Courtesy of Shooters Depot

---

I want to remember it as the spring of 1999 . . . it could have been. I'm getting old timers disease. We were out at what we still refer to as "Shaw's," after famous shooter John Shaw, formally known as the Mid-South Institute of Self Defense Shooting. We were working in the kill houses, doing close quarters battle (CQB) or close quarters combat

(CQC) or whatever you personally might call it . . . room clearing—working the flow of a squad or platoon through a house with live ammo. "A flurry of asskicking," as our chief at the time liked to call it. One of the takedowns we did was completely suppressed. At the time our sidearms were HK Mk 23s, basically the military version of the Heckler and Koch USP .45, with the barrels threaded for suppressors. I loved these pistols. They were total hand cannons, and very accurate. Some of the boys didn't like them, as they were big, and if you didn't have large hands, the weapon could feel unwieldy. With the suppressor attached, you practically wanted to attach a buttstock to it and use it like an SMG, it seemed so big. On these surprise hits, instead of initiating the assault by breaching the door with demo, all is quiet. There is no loud yelling, "Clear right, clear left, all clear." Targets are all taken down by suppressed pistol until your cover is blown and then you go overt. It's the report of the suppressed pistol I remember so well, because even after using it on the range, in the house it just echoed differently. I remember it sounding very much like someone holding a large phone book about three feet off a table and dropping it. It's not that it was anywhere close to "silent," but I tried to put myself in the place of the enemy, perhaps sitting in a back room, not knowing what was coming. The noise—and it was a significant noise—was just random, unidentifiable. So sitting in another room you might perk up, knowing that you heard something, but definitely not associating that something with the crack of a pistol. By then, it's all but over anyway, as a bunch of pipe hitters are about to ruin your day.

From a physics standpoint, the two work on a lot of the same principles, and back in the day people would call suppressors firearm mufflers. There are really two types of suppressors, which are the most common muzzle-based suppressors—what the boys refer to as cans and integral suppressors. Cans are attached to the end of the barrel, and they can be threaded on or mounted in a variety of other ways. Integral suppressors are built right around the barrel of the weapon itself—integrated, right? Both use the same principles.

**An old brochure advertising Hiram Maxim's suppressors**

SHOOTING WITHOUT NOISE

WITH A

MAXIM SILENCER

Made by
THE MAXIM SILENCER COMPANY
Hartford, Conn.

THE TRUE SPORT FOR THE RED BLOODED AMERICAN
Shooting the small bore rifle equipped with a Maxim Silencer is exhilarating sport

HOW THE SILENCER PERFORMS

The Silencer checks the muzzle blast. Instead of the powder gases being liberated into the air instantaneously when the bullet emerges from the muzzle, as in the ordinary gun, the gases are caught by the Silencer.

They are made to whirl around inside the Silencer. This whirling forces the gas to fly out from the center by centrifugal force leaving a central space, just the same as when water is whirled around in a set bowl, a hole or space, forms in the center. This leaves the space for the bullet to make its passage. The gas cannot pass through this space, until it slows down. This causes it to discharge into the atmosphere gradually. This absolutely prevents report noise and also reduces recoil over two-thirds.

As the hole in the Silencer is much larger than the bullet the latter does not touch anything in passing through and consequently accuracy of flight is just the same whether the Silencer is off or on.

HOW MUCH IT REDUCES THE NOISE

The Silencer absolutely annuls all of the noise of the gun report, as can easily be understood from its construction. It also reduces the recoil over two-thirds. The reason for this is that there is a very strong tendency for the Silencer to be blown off the end of the gun. This tendency amounts to a forward pull on the gun barrel. This forward pull counterbalances part of the backward kick of the gun barrel.

The only noise the Silencer does not control is the noise made out in the air beyond the gun by a high velocity bullet in its flight. This noise is a "crack" like the crack of a whip and not a "whistle" or a "shriek" as is commonly supposed. The noise is made in the wake of the bullet and is caused by the same thing that causes the air to crack when a whiplash is snapped.

This noise cannot be avoided when the bullet velocity is high no matter how quiet we make the gun. Where strictly noiseless bullet flight is wanted in addition to a noiseless shooting gun, modified ammunition has to be used.

Its velocity is 1100 feet per second and a heavy bullet is used so as to keep up the striking energy notwithstanding the reduced velocity. It is very accurate up to 250 yards.

The Maxim Silencer is a practical device for use on **REPEATING** and **SINGLE SHOT RIFLES** and on **SINGLE SHOT TARGET PISTOLS.**

Shot cartridges work perfectly in the Silencer and make noiseless miniature trap shooting possible without disturbance or danger. On single shot target pistols it improves marksmanship to a surprising degree as it avoids nearly all recoil.

**Price List of Complete Silencers and Accessories on inside pages**

Photo Credit: Redstick-Firearms.com

Traditional style "can" suppressor. This is a Gemtech HVT model, and is a sweet suppressor to use on both semiautos or bolt-action rifles.

Without nerding out too much, there are two things that cause a gun to make a lot of noise. First, there is the release of gas pressure from the barrel. When the powder from the cartridge explodes, the bullet—which is slightly larger than the barrel so as to engage the rifling grooves—gains speed as it heads for the emergency exit. After the bullet clears the end of the barrel, the pressure from the gas behind the round needs to go somewhere, and that is the first sound you hear as the shooter (not counting the firing pin striking the primer). The

AWC's Ultra II Bolt action integral sound suppressor. AWC uses the latest technology to get match grade quality fully suppressed barrels. Super quiet, and super impressive.

Photo Credit: AWC

second sound is the ballistic crack of a supersonic round as it breaks the sound barrier leaving the barrel (which is approximately 1,100 feet per second, depending on temperature). The last sounds will be either the cycling of the action or the bullet striking a target. As the trigger man, the muzzle blast and ballistic crack essentially become one. If you're downrange though, depending on the weapon and ammunition, you'll most likely hear the ballistic crack before the report of the weapon at the muzzle. That is if you're lucky. If you don't hear the crack while being shot at, most likely you'll have a hole in you somewhere that needs tending to. If using subsonic ammunition, well then obviously this won't come into play.

A suppressor works on a couple of principles. It's all about gas pressure management. First, holes or baffles in the barrel or in the can spill into more space, allowing the gas area to expand before reaching the end of the barrel. The baffles are separated by spacers and employ different patterns and designs to direct the gas to the expansion chambers. A suppressor allows the propellant gas to dissipate, providing more volume that needs to be filled. More volume equals less pressure, hence a lower volume at exit. In addition, the baffles and packing material built into the suppressor allow the gases to cool, and a lower temperature also lowers the pressure. Depending on the model suppressor, some are packed with fiber-based material, so they can be shot "wet." Dunking the suppressor in water prior to shooting will cool the gases down even more, further reducing the sound pressure. Some pistol suppressors are designed to be shot wet and may come prepared with some other liquid in the fiber to aid in the cooling of the gasses, perhaps oil, or some type of gel. These will only last a few rounds and are often unreliable and messy.

**Diagram from a 1971 Heckler and Koch patent. Look to the right of the diagram and notice that initially the barrel is vented out to a wire mesh pack.**

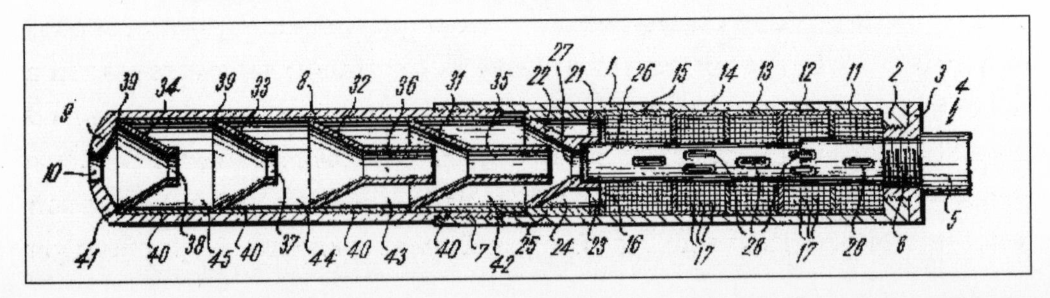

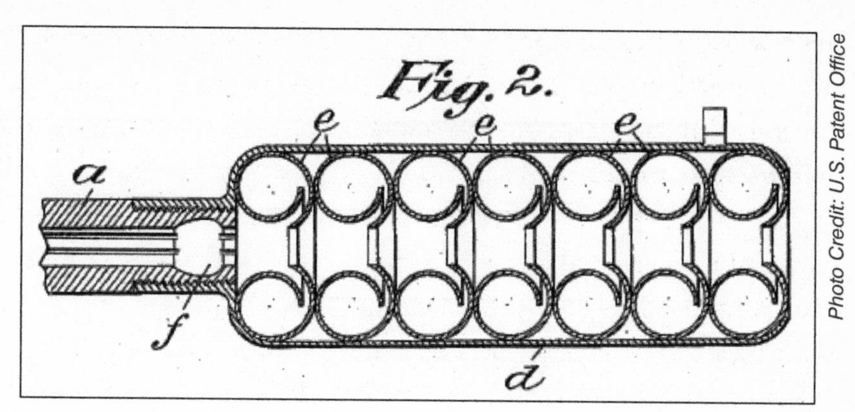


Photo Credit: U.S. Patent Office

**Hiram Maxim's patent drawing for his original suppressor**

Other packing material can be used as well—steel wool, or some other mesh-type material. These will most likely have a longer life span than the fiber but can't be used wet. Last, you're actually giving the propellant gases some space to roam around in, creating traps and turbulence so the gas is delayed and reaches the end of the line slower than it would without the suppressor. By increasing the time the gas has to reach the end of the suppressor, you are changing the energy released, and the actual noise created by the pressure will be different, more of the yellow pages dropping on the table as mentioned earlier than the telltale "BANG" we all know and love.

Most modern suppressors aren't designed to be shot wet anymore and have gone away from any method that will require regular maintenance or take away from the life of the unit. Twenty-first-century models use unique designs to control the gas pressure and also experiment with controlling the actual frequency of the muzzle blast by using phase cancellation and frequency shifting.

Suppressors aren't just for the James Bond moments where you are using a single action .22 for a whisper quiet assassination of some clown in a tux. There are many practical reasons for their use in the field. First and foremost, they protect your hearing, not just in the short term, but in the long term as well. If OSHA had any say, every military weapon would have a suppressor attached. The ability to hear when operating in any situation, the ability to maintain command and control and to keep one of your most important senses keen for situational awareness in a combat situation cannot be stressed enough. Shooting even a 5.56mm round without hearing protection will leave your ears ringing, and get

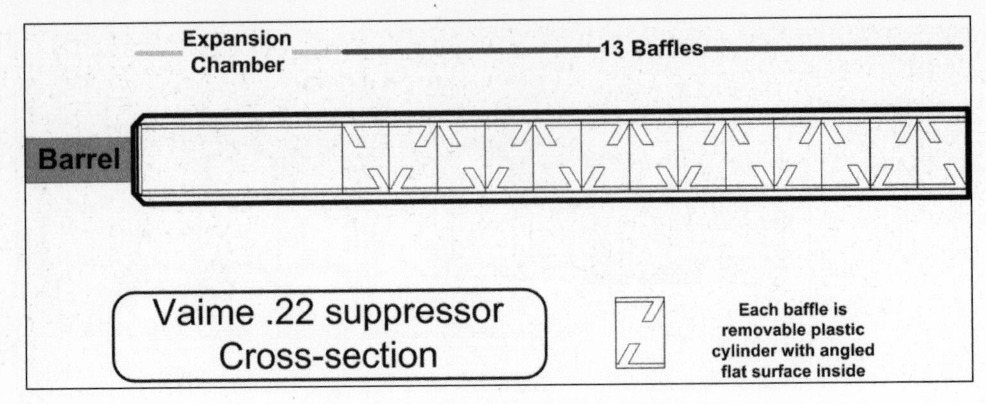


Note the baffles and expansion chamber.

up to the larger calibers and you're doing permanent damage. Take weapons indoors and the noise just seems to keep getting louder, reverberating off the walls. Suppressors were used effectively in Vietnam, taking out sentries, dogs, and other critters used to sound the alarm, and are prevalent today in the global war on terror (GWOT). One suppressed pistol specifically designed for the SEALs in the 1970s was called the "Hush Puppy." I can attest that there are plenty of mongrel dogs in Iraq and Afghanistan that fell to suppressed weapons in order to ensure mission success.

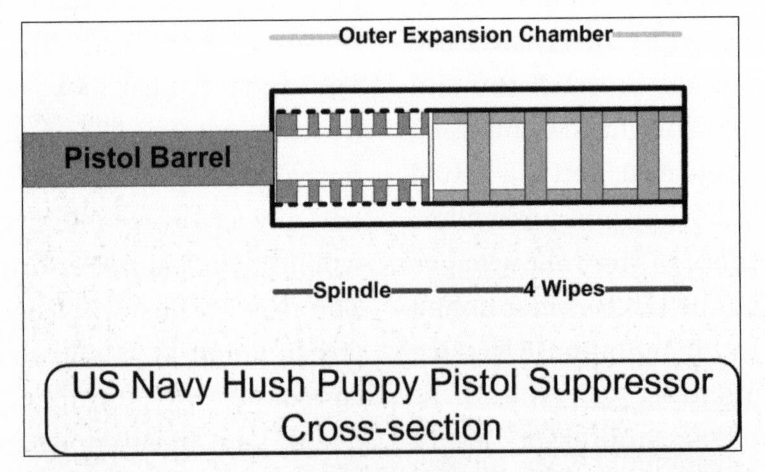


The Hush Puppy was a classic Vietnam-era Smith and Wesson Mk22 Mod 0.9mm equipped with a silencer. You can see from the photo that it uses a "wipe" system, which does exactly what baffles do essentially, creating multiple chambers for the gas to expand, cool, and delay prior to exiting the barrel. The wipes would either be full sheets, be stamped slightly, or have a hole pre-punched for the bullet path. The wipes would need to be replaced frequently, which makes this system relatively obsolete for the modern shooter.

Another of SD Shooters Depot's amazing carbon fiber suppressors. Unlike most suppressors, SD Shooters Depot's patented Advanced Rifle Integral External Suppressor (ARIES) System there is no point of impact change when using the suppressor. Cool.

Using a suppressed sniper rifle at night essentially makes your flash signature all but disappear. Until seeing it with your own eyes, its truly hard to believe how invisible you become. We were training in northern Washington State doing an advanced sniper course at a ranch called Bull Hill in 2002 and did an exercise where half the crew was downrange in the prone position watching the tree line some 400 yards back. In the trees were shooters using the SR-25 with attached Advanced Armament

Coauthor at sunrise in Tikrit with his suppressed SR-25

Corp suppressors, putting rounds on target beyond (and above, obviously) the observers. Individual rounds were fired, and we tried to locate the shooter. It was all but impossible to locate the muzzle flash, but it was good training to hear the telltale "crack" over our heads and also a nice warm and fuzzy feeling when your buddy was pounding a head-sized steel plate from six hundred yards out at night. When using large-caliber rifles, especially when shooting from the prone in a dry or dusty area, a suppressor significantly lowers the amount of dust that kicks up from the muzzle blast, ensuring that you aren't compromising your hide and allowing you to possibly take a follow-on shot if necessary. For any unit doing raids in a crowded neighborhood, hitting a house with suppressed weapons will keep the neighbors wondering what is going on as opposed to being instantly aware that there is a gunfight going on next door. That just amounts to one less thing to worry about. Which is nice.

The recoil of any rifle when using a suppressor will also be substantially lessened, which will allow for a quicker follow-through and acquisition of targets. This cannot be understated, especially with high power rifles like the .338 Lapua and the .50 calibers. If and when you are lucky enough to be able to engage multiple targets in quick succession, lessening the dust cloud kicked up, the sound signature, the muzzle flash (especially at night), and the recoil are all going to lead to a much faster second shot. The dramatic change in the sound emitted after your weapon is fired will also leave the targets confused and unable to pinpoint your location. To quote the Finnish, "A silencer does not make a rifleman silent, but it does make him invisible."

9

# Optics, Ballistic Software, and Other Cool Stuff

Advancement in optics and the integration of technology inside the optics is the way of the future. We have scopes under development right now that have computer and software integrated directly into the scope. The sniper scopes of the 21st century are capable of seeing through smoke, fog, and rain, and can identify targets through buildings at night. They can also adjust for wind and calculate complex ballistic firing solutions that take into account weapon muzzle velocity, environmental factors such as temperature, altitude, humidity, barometric pressure—even how the spin of the Earth (Coriolis effect) will alter the bullet's flight path.

## Rifle Scopes

There are a plethora of great companies producing cutting-edge optics for weapons of the 21st century: Leopold, U.S. Optics, Nightforce, Zeiss, Swarovski, Schmidt and Bender, Bushnell, Burris, and Nikon to name a few of the most prominent. When it comes to choosing optics for your sniper rifle, any one of these companies can provide you with a scope to suit your needs and successfully complete your mission. If you're building your own weapon, consider spending as much as you can afford on good optics, scope rings, and mounts. It will make the difference

## The Coriolis Effect

There's a fair amount of debate on the blogosphere about the relevance of how this phenomenon affects a bullet's point of impact. In layman's terms, you are taking into account the spin of the Earth on its axis when making long-distance shots. Shooting inside 800 yards, you really wouldn't take this into consideration, but when shots are being made beyond 2,000 yards on targets as small as a half-starved Afghani fighter, everything counts. At the poles, the Earth's rotational speed is minimal, but get to the middle latitudes and our little ball is spinning around 800 miles per hour. Head to the equator and its around 1,000 miles per hour. That's moving. In regards to the Coriolis effect, what most people forget to take into account is that it depends on which direction you are shooting—what compass bearing—to determine how much the Earth's rotation is going to change your point of impact. If shooting directly north or south, it will have a much larger effect than if you are shooting due east or west. Let's say that you are shooting a .50 caliber at a compass bearing of 180 at 2,500 yards. With no wind, the untrained will hold for a point of aim, point of impact (all other environmental factors taken out of the equation). The bullet has a muzzle velocity of approximately 2,800 feet per second, and if you're shooting around the midlatitudes, the Earth is spinning at around 1,200 fps. Understandably the Earth, with your target on it, is going to move a little during the 2.5-second flight time, so if you don't account for Coriolis effect, shooting due south, your bullet will miss to the right; if shooting due north you would miss to the left. Savvy? Fortunately I'm Irish so I don't have to worry about any of this. I just get lucky.

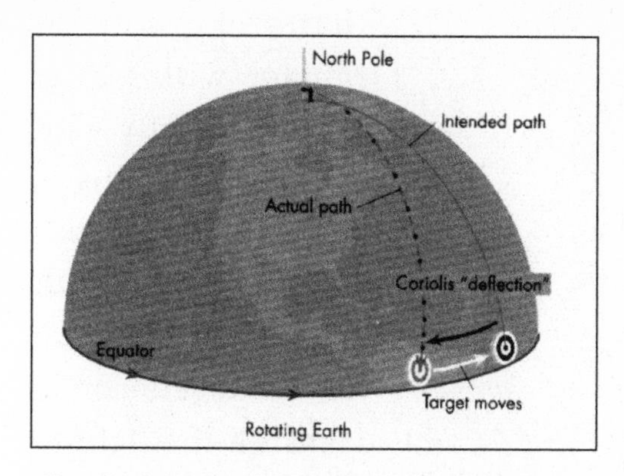

**How the Coriolis effect deviates a projectile from its intended path**

in shots on target at long ranges, reduce eye fatigue, improve clarity, and potentially enhance reliability.

So what the hell does it mean when you buy a 2.5-16 x 42mm scope? It means that you have purchased a 2.5-16x magnification scope with a front (objective) lens diameter of 42mm. Why should that matter? Let's get inside the scope for a second and break it down. Starting from the dangerous end of the rifle, light gathered by the scope moves toward your eye and is gathered up by the objective lens. This lens might be inside a cover, recessed like a scared turtle, and it may also have an anti-reflective coating on it. The cover keeps the glare off and lowers the possibility of a shooter giving away his position, and the coating allows more light to be gathered into the lens and may also have some anti-fogging properties to boot. The bigger the objective lens, the more light that can be brought in. Light is good. Remember that magnification not only makes things bigger, it also reduces the amount of light in the tube, so that at high power you need that big objective lens so that your image isn't blacked out. Four times magnification will quadruple the size of the image; nine is nine times bigger—easy, right? After the image makes its way to the inside, the light waves get bent at the lens, so that somewhere in the tube it's flipped upside down and backward. Because of this, an erector lens, or an erector cell, is used to correct the image so that it appears properly to your eye. With variable power scopes this cell will be contained together in a fixed tube, and this tube moves as a unit as you adjust your magnification. Usually the reticle will be attached to the back portion of these

**Author Brandon Webb going native somewhere along the northern border between Pakistan and Afghanistan.**

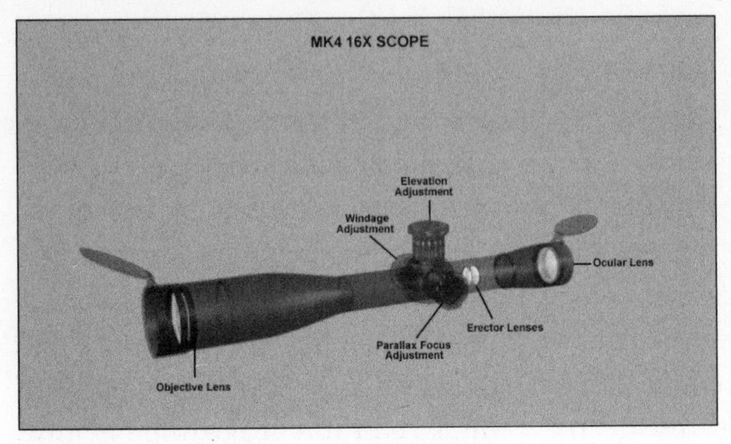


View of Leopold MK4 16x scope

internal lenses. Finally your image comes to you through the ocular lens and provides the exit pupil, which is the size of the column of light that leaves the eyepiece on your scope. The size is determined by dividing the objective lens size by the magnification, so a 5 × 40 will give you an 8 millimeter exit pupil. The smaller the exit pupil, the closer your eye will have to be to the ocular lens for a proper view. Too small and you'll be right up on it, getting smacked in the face every time you pull the trigger. The size of the exit pupil is also important as it gives you more wiggle room when finding the appropriate eye relief, which is the point at which all the scope shadow disappears, and you are looking at a full tube's view of a crystal clear, bright image. Adjusting windage and elevation moves the internal lens assembly in relation to the outer lenses, so that your point of impact will adjust ever so slightly with those clicks. This is why the nicer optics will provide .25 MOA adjustments. So lenses that efficiently transmit the light to your eye will provide you with the brightness that you want, and high quality properly ground lens glass will give you the clarity you need to properly ID your target.

But even if you have access to the most expensive sights, if you don't how to properly mount them to your weapon and set them up for your eye relief and body size then you may as well stick with iron sights. The

Mil Dot reticles can be illuminated for night shooting.

authors have taught several long-range shooting courses and have seen well-heeled civilian shooters show up with the latest Accuracy International .308 and a top-of-the-line Leopold Mk IV scope but complain that they just can't seem to get any consistency out of their groups. Usually, a quick shake of the scope itself will find it loose, with the mounting bolts not even tightened down. As the old adage goes: "If you're going to be stupid, you'd better be tough." Other common setup errors that we saw were setting natural eye relief and failing to set the reticle in a perfect plane relative to the weapon. Before doing anything else, you have to find the appropriate eye relief. Start with the scope toward the dangerous end of the weapon and get in a comfortable prone shooting position with your eyes closed. When settled, open your eyes, and move the scope into a position where scope shadow is eliminated. Try other shooting positions to ensure the location is correct, and then lock those bolts down. Essentially, for a traditional Mil-Dot scope, when on target the reticle should look like a "T" not an "X." Simply loosening the scope rings and rotating the scope will quickly fix that. It's hard to have an accurate wind hold and stay on target when your reticle is in the X position . . . your vertical point of impact will take a beating! Next up is focusing the reticle. If it weren't for the fact that most men hate reading instructions and inherently know how to do everything, we wouldn't see so many of these problems on the range. In order to get the reticle crisp, point the scope at a white background, and rotate the *eyepiece* until it is in perfect focus for *your* eye. If your scope has an eyepiece lock ring, after getting the reticle focus dialed lock that baby down, too. Understand that as you get older and your eyesight deteriorates, you may need to readjust this somewhere down the road.

**Coauthor Glen Doherty keeps warm with some aftermarket ghillie product. Don't forget to cammie and veg up your scope and weapon.**

*Photo courtesy of NightForce*

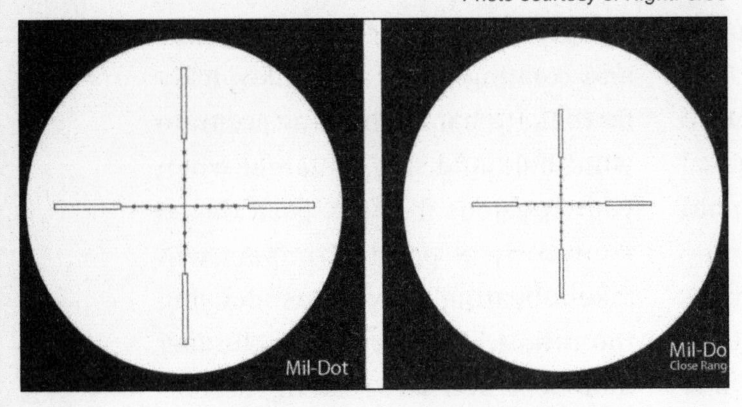


**Two Nightforce reticles using the classic Mil-Dot system that military shooters are trained on**

*Photo courtesy of NightForce*

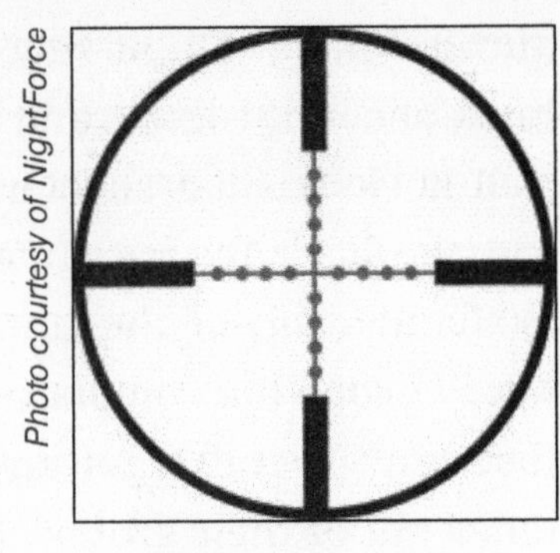

**Mil Dot reticles can be illuminated for night shooting.**

Currently in use and one of the most popular scoped optics employed by the Special Operations community is the variable power Nightforce with the illuminated reticle. The variable power allows for a lot of flexibility between rural and urban operations and the illuminated reticle enables the sniper to operate at night in an urban setting where ambient light is available without night vision. There have been many Al Qaeda operatives who have fallen prey to the Nightforce scope and that illuminated reticle. While superior optics exist, you cannot beat the Nightforce for price and reliability.

## What is Parallax?

The majority of scopes used for long range shooting will have three knobs on them in a cluster in the middle of the unit. Two are obvious: your windage and elevation knobs. But what about that third knob with the infinity symbol on the left side of your scope? That is your parallax adjustment. So what is parallax? By definition parallax is the apparent displacement or the difference in apparent direction of an object as seen from two different points not on a straight line with the object. OK, but what does that mean? It means that it adjusts the internal focal planes of your scope so that the reticle and object image are on the same optical

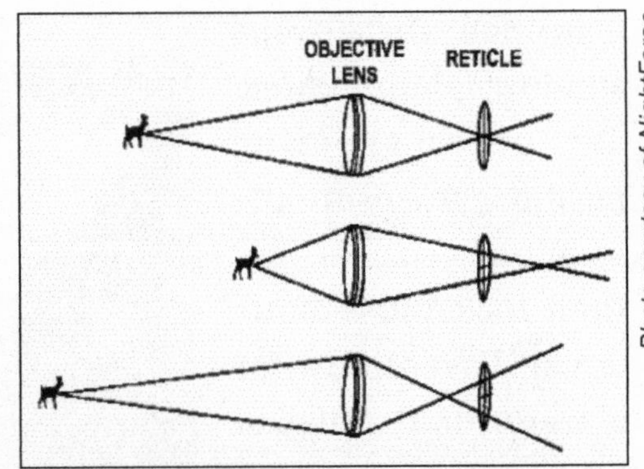


*Photo courtesy of NightForce*

**The top image shows no parallax. In the lower two, you can see that the reticle plane is either in front of or behind the object plane. No bueno.**

planes. If you have parallax that little 'picture' of your target is being transmitted either in front of or behind the reticle. Adjusting the knob will move that image plane so that it falls exactly where its supposed to.

To illustrate the basic principle, imagine that you are the passenger riding in a car and you look over at the speedometer and it appears to you to read 80 mph. (Get some!) But the actual speed and what the driver sees is 100 mph. The way your eye lines up with the needle (which represents the reticle in this scenario) and the numbers on the instrument panel (the object/target) is seen from a different angle from where you are sitting. This isn't exactly accurate, but you are getting the idea, right? Now try holding your index finger up at arm's length in front of you, closing your left eye and 'sighting in' on something on the wall. Everything is perfectly lined up. Now close your right eye and open your left. Your 'reticle' (finger) has jumped off target. Your image of your target and your reticle were not in the same plane, so you have parallax. With scopes, especially high powered scopes, this can be a serious problem if not remedied. A lot of people you ask will tell you that the dial that adjusts for parallax is actually an 'object focus' or often a 'reticle focus' knob. Not so much, although sometimes the parallax adjustment does make the object go in and out of focus, especially at long ranges.

The parallax adjustment knob will usually have distances stamped into it, but they are rarely dead on. The way to check for parallax is to get in a nice stable position, and to move your eye/head slightly up and down or side to side. If the reticle, or the target, appear to move, then you have parallax and need to adjust until that movement does not exist. Remember this though: parallax increases with magnification, so if

*Photo courtesy of U.S. Optics*

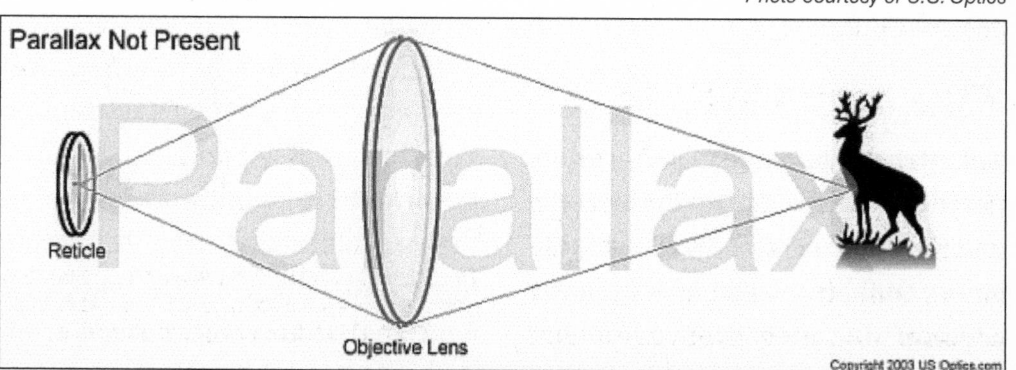

you were shooting at 100 yards and had forgotten to adjust for parallax at that range, you were maybe .1mm off inside your scope, and then had to quickly send a round downrange at 1000 yards so you could power your magnification up 15x on your variable power scope. Your image plane will now be 1.5mm off the reticle plane, assuming that you were shooting at 1x magnification initially. Too nerdy for you? So long as you've come away with an understanding that the knob isn't for adjusting focus, we've done our job here. Good luck.

## Spotting Scopes

Modern snipers are going to variable power scopes because they are much more practical at a variety of ranges. You can dial power back in an urban setting or dial power out in a long-range rural environment. Having a high power setting also lets the sniper spot for himself using the increased magnification to get positive target identification or a read on environmental conditions at the target.

Most long-range tactical rifle scopes will offer you variable magnification, typically 3.5x–10x, but on the extreme end up to 8.5x–25x. The variable power enables the scope to be more practical in a variety of environments. That high power magnification isn't commonly seen on the weapons of choice used by today's operators in most environments, but having the ability to really get out there and get a good look at a target is why snipers, especially when operating as a two-man team, will bring a high powered spotting scope with them. These spotting scopes will usually provide magnification starting around 20 times, and going all the way to 100 times on some of the more powerful models. Most commonly you would find a 20–60x model slid into the snipers drag bag. Let's not forget that one of the sniper's primary jobs is trained observer. His ability to stay hidden and to observe a target and relay pertinent information back to command can make the difference in any engagement. Modern spotting scopes can be hooked up to digital cameras and video recorders so that 100 times magnification can be used for positive identification of HVTs (high value targets) or to prepare a target package for a follow-on direct action assault of a hard target. Still photos or digital video can then be saved, compressed, and sent via satellite anywhere in the world. I'm sure the president and the joint chiefs have had coffee while watching near real-time video passed from

Leica Rangemaster Laser Rangefinder. Rugged and incredibly handy, it can accurately range in almost any weather conditions out to 1,200 yards.

snipers observing a target in the field on the other side of the world. Not to mention, it can be fun to be able to review your long successful shots again and again and show off to your friends. (If you haven't already done so, go to YouTube and check out the video taken through a spotting scope of Canadian snipers taking out Taliban at over 2,000 yards with a .50 caliber. Awesome.)

The additional magnification is also useful in training on the range and calling accurate wind for your shooting partner. Especially when the weather and wind are up, "seeing" wind and mirage is easier with the nicer glass. Then, of course, there's watching "trace." There are long-time shooters that claim to never have seen it, but get on a nice spotting scope behind a shooter on a warm day and you get to have your *Matrix* moment. You can see the bullet's pressure wave as it tracks through the air and can follow it all the way into the target. This is where the shooter/spotter pair combination shines. The shooter will "call his shot," letting the spotter know exactly what his sight picture was when the surprise break happened, and having actually seen where the bullet impacted from the trace or target splash, the spotter can make the necessary corrections

Leica spotting scope fitted with a digital camera. This was a favorite among SEAL sniper instructors.

Leica pocket laser range finder. This is an excellent battlefield-ready range finder.

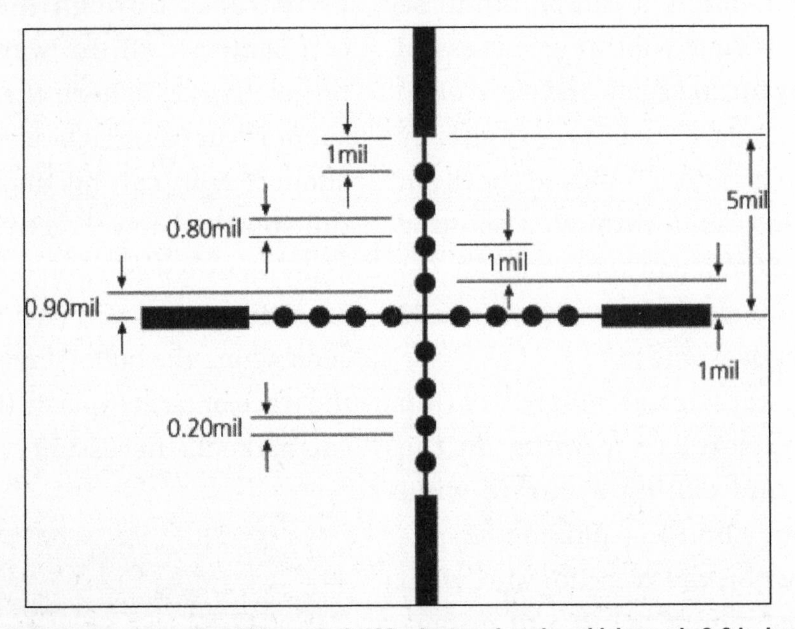


The Mil breakdown on a scope...One Mil equals 3.438 minutes of angle, which equals 3.6 inches at 100 yards. For ranging purposes, you have to know or approximate the height (or width) of your target. The formula is: Height of Target (yards) x 1000/Height of Target (Mils)=Range in Yards. There are tables that can help with quick ranging, as well as useful tools with sliding scales like the Mil Dot Master. It's always good to know how to do it the old school way just in case technology lets you down. Batteries don't last forever. The modern method would of course involve using a laser rangefinder. These can be monocular or purchased integrated into binoculars like Vectronixs' Vector which we used in the field in the SEAL teams. The Vector has 7x magnification, and also has a built in digital magnetic compass, a clinometer for all important angle shooting adjustments, and data links so the information and images can be downloaded if desired. Cool, right? We always thought so.

for windage and elevation to ensure that any follow-on shots are landing exactly where the pair wants them.

## Laser Range Finders

The future will integrate laser range finding capabilities directly into your rifles optics. In fact, it's happening now, but not to the level and quality you need when getting out past 1,000 yards. Accurately ranging your target is key, and there are many ways to do it. The Mil Dot reticle scope is a powerful measurement tool that lets a trained shooter quickly and accurately range for distance and hold for leads by understanding that minutes of angle are visually

### Story from Brandon:

"We were testing some ballistic software applications when I was attached to the Naval Special Warfare Group One sniper cell. As part of our advanced sniper sustainment training we would take the guys on several hunting trips each year to stay on top of their game. Stopping a beating heart in an animal is very similar to a human and it's good practice. On this trip we were hunting white-tailed deer on the western border of Canada and I had a brand new .300 Win Mag that was new out of the box. The guys were slow to trust technology so I decided to do a test run of the ability to sight in a rifle in one environment and account for a totally different environment with the help of ballistic software. I took the gun to one of the sniper ranges at Camp Pendleton and shot a three round zero at 100 yards and made the necessary adjustments to my Nightforce scope. At the same time, I shot my zero through a chronograph to measure my .300's muzzle velocity (it was averaging 2,850 feet per second). I then went home and looked up the weather conditions for the remote lodge I would be at in a few days and entered the data into my program. I programmed it to give me my come ups out to 1,000 yards. Three rounds through a brand new rifle system and I was prepared to kill out to 1,000 yards! You have to love technology. A few days later I set out into the mountains in a light rain to still hunt. Deer are like humans and don't like to get wet and I decided to use this to my advantage as I was counting on the forecast that the rain would let up in time for the deer to come out and feed before sunset. After about an hour hike, I had a great elevated vantage point of a small field where I had glassed trace with my binos. The center of the field ranged out at about 460 yards. As soon as the rain stopped, there was a mad rush of hungry white-tailed deer and soon I had a nice-sized buck in my scope. I had already dialed in my elevation for 460 yards and squeezed off a great shot that lifted the deer off its feet and killed it instantly, point of aim, point of impact right through the heart. From that moment on I was an evangelist for applying and leveraging technology to the sniper's advantage. I had taken a brand new rifle, sighted it in California with three shots, took it to high elevation and a very cold climate (a radically different environment), plugged in my new environmental conditions (temperature, barometric pressure, type of round, and degree of latitude), and with the fourth shot out of the gun I had executed a perfect kill shot."

**Simple reticle on a 4x scope. Sometimes simple is all you need.**

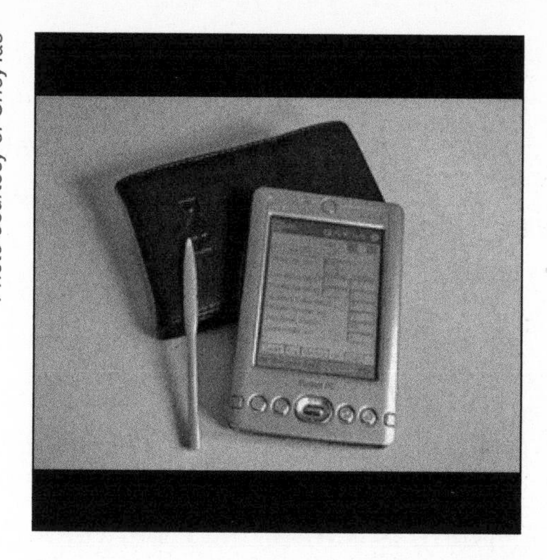

*Photo courtesy of CheyTac*

**CheyTac developed a very accurate ballistic program after integrating radar data gathered on bullet flights at the Yuma Proving Grounds. One only needs to input a few variables to get firing solutions out past 2000 yards and achieve first round hits. Their software shown here in a pocket PC, but can be used in a variety of handheld devices.**

*Photo courtesy of Knights Armament*

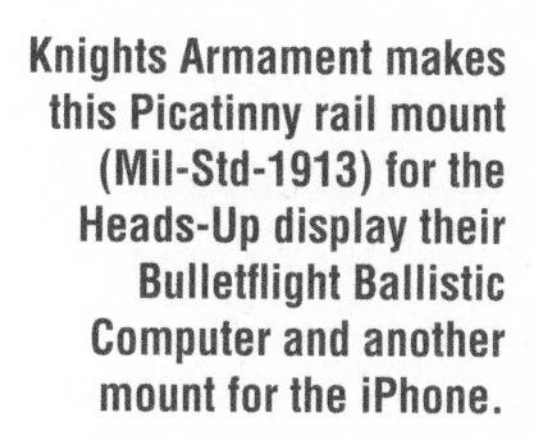

**Knights Armament makes this Picatinny rail mount (Mil-Std-1913) for the Heads-Up display their Bulletflight Ballistic Computer and another mount for the iPhone.**

represented through the dots and spaces between the dots on the scope. Any good sniper will know and have his Mil Dot leads down cold. The advantage of this reticle versus traditional scope sights is that a sniper can adjust quickly for multiple targets and account for a variety of changing conditions.

Modern day ballistic software gives the 21st-century sniper a huge advantage. A single sniper using a small hand held computer, ballistic software and a chronograph, can sight in multiple rifles at the 100 yard line with only a few shots per rifle. With the advantage of software and access to the internet (to obtain environmental conditions and elevation), the modern sniper can be prepared to go anywhere in the world, take his rifle out of the box ready to kill. Cool stuff if you ask us.

## A Quick Aside on Ammunition and Ballistics:

So many different types of ammunition are in use in the world today. In the study of the history of precision marksmanship we've evolved from the smooth bore .50 caliber round ball to custom made tri-metal blends boasting over 5,000 feet per second out of the muzzle. Discussing ballistics to the layman and even to a shooter who considers themselves experienced is often enlightening. There is no art to ballistics, but only science, and in understanding the physics throughout the firing and impact cycle will make everyone a better and more well-rounded shooter. It is essential to have a basic understanding of gyroscopic stability, wind drift, rifling, muzzle velocity, trajectory, Coreolis effect, and terminal ballistics (what happens after the bullet strikes its target). There is a difference between the ammunition that the regular ground soldier is using and the rounds the sniper is equipped with, which is often considered "match grade." What's the difference? Why is this important? It's all about accuracy and first round effectiveness. Not that this

Ballistic Software developed specifically for the iPhone. Amazing.

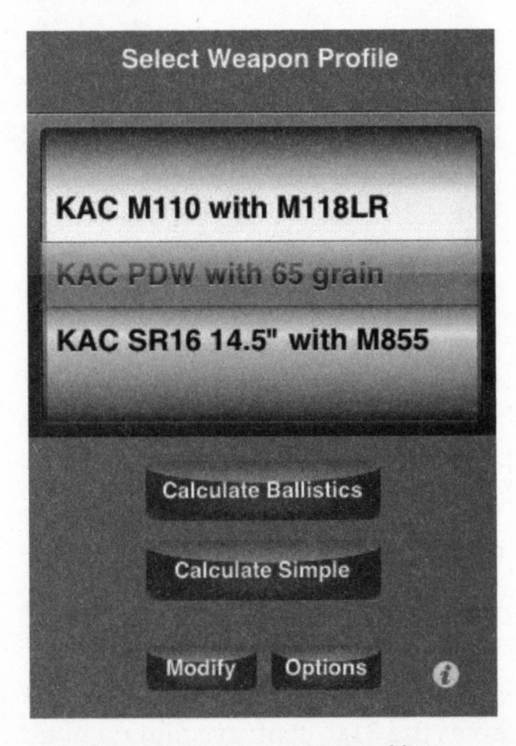


Select your weapon and ammunition.

**Input environmentals and get your firing solution. Almost takes the fun out of it. Almost.**

*Photo courtesy of Barrett*

**Barrett's Barrett Optical Ranging System (BORS) Ballistic Computer mounts directly on your scope and couples directly to your elevation knob. BORS can be set for your specific round, and will automatically compensate for environmentals, distance, and even shooting angle. In the future, expect to see this technology not as an add-on to a scope, but built directly into the scope itself.**

**This Kestrel 4500 is an anemometer (wind reader) and so much more. It can quickly and accurately give you just about every environmental stat you need: barometric pressure, altitude, humidity, temperature, wind speed, digital compass heading, even wind direction. A handy tool when needing data quickly for your ballistic computer.**

Photo Credit: Tom Gaylord

**An Alpha Chrony from Shooting Chrony. A chronograph will measure the actual velocity of whatever round you are using, and is an important tool in any 21st-century sniper's kit. If using factory ammunition or reloading, getting a true muzzle velocity for any given round is critical information to input into your ballistic software to produce an accurate firing solution. Just because the manufacturer tells you that their ammo delivers 2780 feet per second, doesn't make it so. It's best to check. Consistency is paramount in shooting, and finding and using the same lot or batch of ammunition is important to maintain consistent rounds on target, especially when really reaching out to touch someone.**

isn't important to every soldier, but there is also a cost involved. Match grade ammunition, ammo that is made to a higher standard, costs more, and is less readily available. Attend any long distance shooting event in the world and most of the small talk you will hear at the firing line is about types of powder, custom casings, bullet weights, tricks, myths, and black magic. To those who know, it's the make or break difference, the difference between winning and losing a National Shooting Championship. To the majority of military snipers, it follows more along the lines of "Who cares?" as custom ammunition isn't part of our world. This isn't to question the relevancy or authenticity of the hand-loads and their effectiveness; it is just simply the facts. We shoot what we are given, and still achieve grand champion-like results.

Burris optics has created a unique system they call the Eliminator laser scope. It is preprogrammed for more than 600 commercial cartridges, and unlike all the other systems we've presented in this chapter, the Eliminator is completely self contained. The scope is a 4-12x-42mm, very respectable in terms of magnification with an objective lens diameter of 42mm. Its laser rangefinder won't reach out as far as some of the powerful handhelds; it's only able to reflect off a human sized target to 550 yards. Not yet anyway. A couple years down the road, more companies

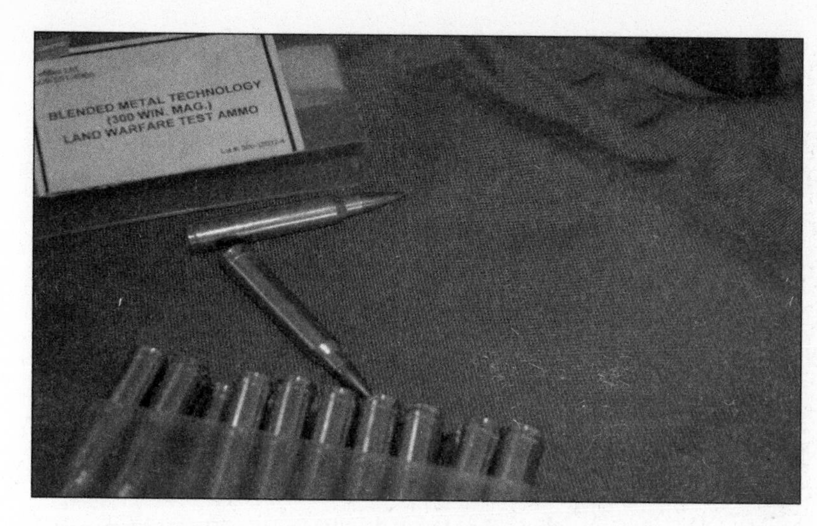


21st-century ammo: LeMas Ltd. is producing some cutting edge ammunition with incredible results. The authors had the opportunity to demo some of their .300 WinMag and were blown away by the ballistics, especially at the terminal end. Way above and beyond the stopping power of a traditional .300 round.

*Photo courtesy of Burris.*

The Burris optics"Eliminator" laser scope.

*Photo courtesy of Burris*

Burris Eliminator. Keeps you from having to bring too much shit along with you . . . which is nice.

will be following up Burris' lead by integrating everything into one user friendly scope—rangefinder, night-vision, thermal, laser designators, ballistic computers, and more. Burris asks that you put in a simple ballistic figure: your bullet drop at 500 yards when zeroed at 100, then verify at 500 yards. From that point forward, all you do is range your target, then the scope will calculate for the angle of the shot, the ballistics profile of your particular cartridge, and put a lighted dot on the vertical post of the reticle where your holdover is. (It does not miraculously hold for wind . . . again, not yet.)

## Night Vison Optics

"We own the night" is an expression commonly heard when getting tales from overseas. It's so true. From driving to a target you are about to hit totally blacked out, to being able to glass and scan areas completely on night vision creates an incredible advantage to the operator. Typically, the majority of the platoon/squad will be on helmet mounted NODs (Night Observation Device) such as the AN/PVS-7 which is common double-eye night vision, or perhaps the PVS-14 Monocular Night Vision Device which is better if you're on foot and doing long patrols, as it allows you to continue to use your natural night vision and depth perception to help keep you from falling on your face. These will be coupled with laser systems mounted on the Picatinny rails of your primary weapon like the PEQ-2, which is an IR target pointer and aiming light, coupled with an infrared illuminator . . . think Surefire flashlight but not visible to the naked eye. Insight Technology recently put out the AN/PSQ-23 STORM (small tactical optical rifle mounted) which also has a laser range finder and built in digital compass in addition to all the features of the older PEQ-2.

Infrared lasers and visible lasers are good in certain situations (think closer range and urban operations), so for those kind of missions you might find a sniper having his weapons system equipped accordingly. But if you are looking to reach out, have a lot of ground to cover, and want surgical precision there are much better setups available, and the 21st century has solved a lot of technological problems that plagued snipers back in the 90's. Today a shooter has a plenty of choices, starting with easily attached models like the KN203FAB MK IV or the KN-250, both manufactured by different companies but providing the same type of proven solution. The

**Nice AR shown here with the PEQ-2 mounted on the top Picatinny rail, with a touch pad switch secured by velcro on the right side of the forward hand guard. With the Trijicon ACOG sight, this kit will get the job done in a lot of situations.**

systems slip into the front of whatever day scope is already mounted, and after a quick boresight on the unit (which is easily done in the field), the weapon system is ready for night operations. You could also flip the front cover on the unit down and leave it in place for daytime operations as well, but it tends to be a bit unwieldy. It's better to stow it until the sun starts going down. These image intensifiers allow you to see your regular reticle, and use all the functions of your day scope without putting a whole new optical unit on there, and that is invaluable when snipers rely on consistency and routine prior to taking difficult shots. Just like a pro golfer, there is a program that is undertaken before every good shot . . . it's all about the setup, right? Get your mental and physical plans on the same page, envision a hit, and then go through your pre-shot routine.

*Photo courtesy of CheyTac.*

**SPA-Defense (Simrad) couples their KN203FAB Night weapon sights to these CheyTac sniper rifles. Bad guys beware—no more hiding in dark corners. You can see though, that at 3.5 pounds, they would add a top heavy component to the weapon system. So long as you are in a good supported position though, there are no worries.**

Accuracy International 338 rocking the PVS 26. This setup is fantastic.

If your weapon is fortunate enough to have the full MIL-STD 1913 rail interface (also known as the Picatinny rail), then there are definitely other options available. There are multiple systems that will also mate with the front of your day scope, but have their own rail mount system so that they are locked in. Some prime examples are the AN/PVS-22 Universal Night Sight, the PVS-24, or the PVS-26 which is essentially the same as the 22 but specifically designed for long range. These units attache to the rail in front of your optics, and add night vision capabilities with Generation 3 image intensifiers (which gather and increase ambient light) and can be installed and removed without any tools.

Another added bonus to a Universal Night Sight (UNS) is that there are no adjustments to be made to get it dialed into your weapon. It has permanent boresight alignment, and that will give the shooter a nice warm and fuzzy feeling, especially if a quick shot needs to be made and there was a lot of maneuvering to get into a good firing position. Another reason the UNS is a good item to keep in your kit is that it can also be coupled to a spotting scope if you really need to identify targets in a low light environment. Additionally, it can be used as a handheld monocular device. Sometimes that's really all you need—K.I.S.S. (keep it simple stupid). The newer version of the PVS-22 is called the MUNS (Magnum Universal Night Sight). It operates on all the same principles, but is just the latest and greatest (and most expensive) version. It gathers nearly twice the light as the 22, and

Photo courtesy of McCann Industries

**AN/PVS-22 UNS seen coupled with a Leopold MK 4 Spotting Scope with the McCann Industries Optical Mounting System (MOMS).**

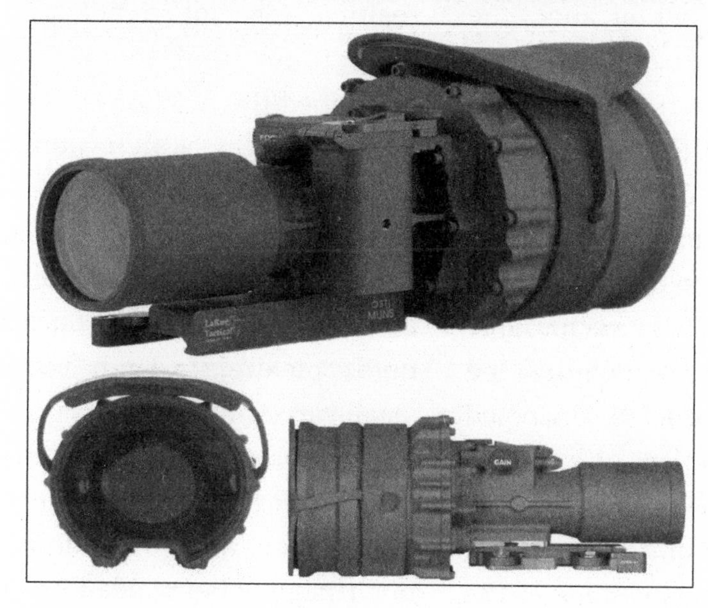

**Optical Systems Technology makes both the MUNS and the PVS-22 UNS sights seen here.**

can engage targets approximately 1.5 times farther than the original UNS. Break out your wallet Uncle Sam, the boys want the best.

There are thousands of these in-line night vision devices deployed overseas right now, and thery are an essential addition to any 21st-century sniper's war chest. Believe us when we tell you that these sights are getting it done right now. Tonight!

## Thermal Optics

Let's take it to the next level and talk thermal. The 21 century just keeps pushing the technological envelope, and the latest thermal imaging/ Forward Looking Infrared(FLIR) sights do not disappoint. Infrared energy, what we'll call heat here, is invisible to the human eye, but these systems enable us to "see" the heat emitted from any object, both animate and inanimate. To break out a bit of science, at the lower end of the visible light spectrum is the color red, the upper end is violet. When you hear people refer to the ever fearful UV rays, yes, they are talking about the light above the color violet that sports a shorter wavelength and harbors more energy—so we get ultra-violet. On the other end of the spectrum, just below red, is infra-red, and it's here that this technology lives. There are two ways that this technology makes it to the markets: uncooled and cryogenically cooled. Uncooled systems are the ones the operators in the field are using every day. They are less expensive and more durable, but they are also not as awesome as the cryogenically cooled units. These bad boys innards are sealed in a system that keeps them at a temperature below 32 °F degrees Fahrenheit. Operating at the low temperatures allows the sensors to produce incredibly clear images, detecting as small as a .2 °F change from over 300 meters. Still, these are too fragile to take into the field. For now. It's a competitive market out there, and several companies are leading the way, making what was once a heavy, bulky, power sucking device into something rugged, manageable and able to last in the field. Some of the newer (and ridiculously expensive) devices actually combine image intensifiers (night vision) with thermal, collect the two images, then overlay them digitally into your field of view realtime. Amazing, right?

Raytheon and the U.S. Army developed the AN/PAS-13 Thermal Weapon Sight, which along with other companies FLIR/thermal systems offers some distinct advantages over regular night vision. First and foremost, it operates day *or* night, picking up even extremely small variations in temperature, and will not be overly affected by rain, smoke or fog. Image intensifiers need some light to work, but thermal does not. (Imagine being deep into triple canopy jungle. Nothing penetrates that. The night there is like being deep in a cave. In that scenario, hope you brought your thermal goggles.) There are different viewing modes on most thermal devices, enabling the user to view in "white hot" or "black hot" modes. It's like playing around with picture editing software, or reversing a negative—sometimes it's just easier for the eye to recognize a shape or identify a particular target in white or in black. Either way, it's cool. Additionally, thermal sights are unaffected when hit by direct high intensity light which would cause night vision to shut down. The night vision doesn't exactly shut down, but you do when on NODS and hit right in the lens with a powerful overt light or if someone or something pops off flares. It's called "bloom", and it will temporarily blind you. The PAS-13 comes in a variety of different sizes and strengths: light, medium, and heavy. The light version is for your rifle or carbine and can positively detect targets out just past 500 yards or so. The medium you would see on the Squad Automatic Weapon (SAW) or the M240 (the new M60 designator . . . 7.62 Automatic weapon, think Rambo), and the heavy on the up guns like the .50 cal. Each size weighs a bit more, but also increases the effective range and battery life, so pick one appropriate for the equipment, and go get some.

There are even smaller thermal sights on the market, and some that work just like the UNS discussed earlier. Insight technologies has created the very small and light CNVD-T clip-on thermal weapons sight, which can be used as a stand-alone day or night sight, or be mounted in front of your day scope without affecting the day scopes point of aim or impact. For snipers, the ability to have one of these small units tucked into his dragbag is invaluable not just for identifying targets in adverse light and weather conditions, but for reaching out and touching them as well. The future is going to bring even more incredible technology to the frontlines and to all shooters. Keep your eyes open, because we are all going to witness amazing things to come: imagine an auto ranging 2x-40x lightweight sight with overlaid thermal and night vision capabilities, built-in visible and IR lasers along with IR floodlights and overt white light, integrated ballistics computers that calculate wind, elevation, angle, envi-

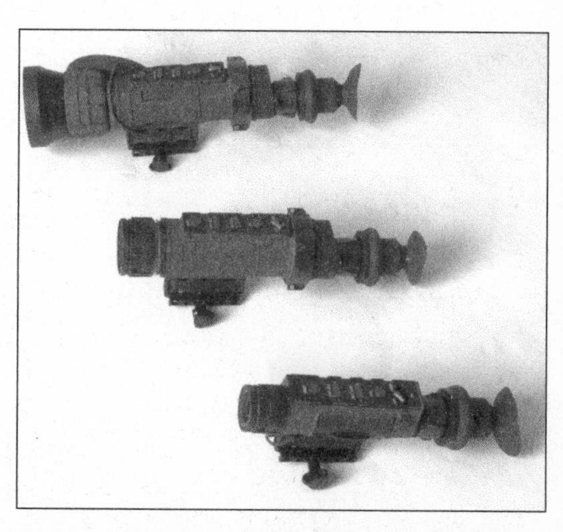

**Your collection of PAS-13's. From top to bottom: heavy, medium, and light.**

collapsible weapon that fires smart bullets—bullets that turn corners, track on laser designated targets, and offer a user selected terminal ballistic effect. (Think penetration, or explosion, or *both.*) Well, a kid's got to dream, right?

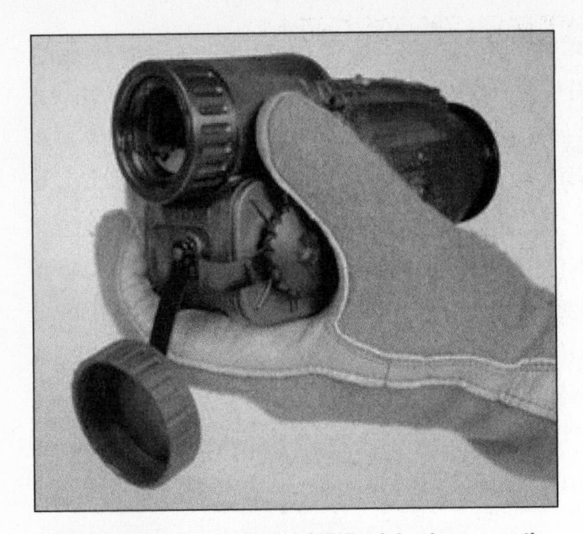

**The small and very portable CNVD-t (also known as the SU-232/PAS) Clip-on Thermal Device**

ronmental factors, and are attached to a lightweight mostly carbon fiber

---

"We were in the Philippines conducting a Foreign Internal Defense (FID) mission, training a couple platoons of Filipino SEAL's for 8 weeks. These are great missions on more than one level. Being in Subic Bay living on the beach with no other U.S. military around for starters. The diving in the bay is amazing—warm water, and a plethora of scuttled U.S. Navy ships at reachable depths left behind after the U.S. pulled out of the P.I. in World War II. A casino down the street, amazing tropical weather, and great cheap food. Besides the liberty, the training we put on for the Filipino SEALs is top notch. They have modeled their SEALs after our program, but they just don't have access to the level of instruction we do, nor the equipment, training facilities, or food. (you should see how thin these guys were.) These guys are hard though, and the majority had seen plenty of combat in the jungle of the Philippines chasing Abu Sayef and other terrorist organizations

**Coauthor Glen Doherty selling his soul for a cheesy photo op in the Philippines. This was one of our little makeshift huts in the village that the Filipino SEALs did a Direct Action (DA) assault on.**

around. Toward the end of our stay, we had them do a Final Training Exercise (FTX), a diverse mission profile that would take several days and force them to use the majority of their skills. The mission was to send in a small Surveillance and Reconnaissance team to watch a village out in the jungle for a couple days, send back detailed imagery of activity taking place there and identify High Value Targets (HVTs)—we were the high value targets. The recon team inserted by boat, and did a long swim to a remote beach, then slipped up into the hinterland and worked their way just below a high ridge some distance off from the village. There were several of us glassing the jungle for movement for hours, trying to locate and bust the team coming in. Simple fact being these guys moved in the jungle well . . . better than we did . . . most had grown up in it. We had given up trying to find the team with Night Vision and Binos—it just wasn't happening—but then we did happen to have a PAS-13 with us. It might have been 10 seconds after we turned it on and started scanning the ridge that we picked up four perfect images of our boys up there, some 800 meters off. If ever there was a proof of concept for technology, this was it. Good times in the P.I."

# 10

# Lasers

Military use of lasers is growing in the urban warfare theatres around the world. They offer a unique sighting and aiming ability that can also have a tremendous psychological effect on enemy combatants.

Laser sites were first made practical with the invention of the laser diode; prior to that the power requirements and general fragility kept lasers confined to larger deployments (trucks, tanks, aircraft, ships, etc.). With the advent of the semi conductor laser in 1968 and subsequent miniaturization, the laser sight became real, no longer science fiction; it still was not the death ray, but then, we don't have flying saucers yet either.

The first laser sights were red in color, poorly collimated (meaning the laser beam would break up after a short distance), and ate batteries like pizza at a Weight Watchers meeting. The beam was nearly round in most cases and bright (laser sights are several times brighter than the sun at their operating color and frequency). Now they come in red, green, blue, and infrared (useful for aiming with night vision).

The first infrared solid state lasers were used in laser printers, were highly collimated, and had low power

"I remember how effective my visible laser was to me while conducting sniper over watch on the USS *Cole* after the suicide bombers killed over a dozen crew and nearly sank the ship. I was set up on the bridge and built a hasty hide to blend in with the ship's superstructure. I had a .50 caliber and three law rockets and orders to shoot to kill anything or person that compromised the perimeter. Not bad ROEs if you ask me. The situation in Aden was tense and we were surrounded by hostiles that had major weaponry trained on the ship. At night I shifted positions and started using my high-powered visible laser to scare the Yemeni weapons crews. After a few sessions of Lasik surgery, all crews shifted their weapons off the *Cole*. This brought a smile to my face and made the situation a little less tense. I don't care who you are, no one likes guns pointed at them."

requirements. They proved an excellent cross-over technology into the military domain. With night vision scarce on the battle field initially, their application wasn't very practical, but over time the IR laser targeting illuminator has become the go-to choice for an "invisible" laser weapons sight.

Laser technology is used to range, illuminate, target, and mark targets for following on Close Air Support (CAS) missions. Infrared laser systems add to our ability in certain environments to "own the night." Visible lasers, while known to the general public as a favorite of

Laser Devices Inc.'s DBAL-A2 is a dual beam aiming laser which has both IR and visible laser pointers in one unit, combined with a focusable infrared laser illuminator (think giant flashlight when on nightvision). The laser targeting is fully adjustable for windage and elevation, but at around 1MOA adjustments, this unit is not designed to go out to extreme distances. In close quarter urban environments, however, enemy beware.

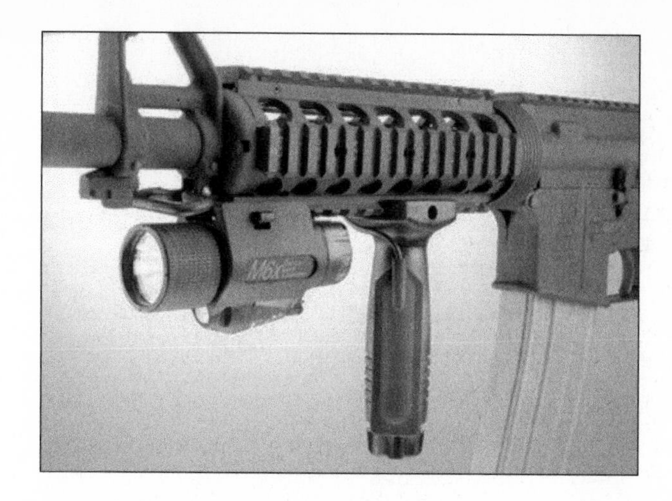


The Insight M6X Laser Tactical Illuminator flashlight shown mounted on the Picatinny rail system. This one is also showing the touch pad activation switch on the vertical grip, which makes for easy on, easy off. During the 20th century, you would have had two separate units to accomplish what this much smaller system can.

science-fiction and Hollywood, do have some use in combat. They do offer a unique sighting and aiming application, but can also be used as psychological warfare in situations that might not have escalated to the point of no return. No matter who you are, or where you are in the world, seeing an illuminated dot holding on your chest will cause you to rethink any mischief you had in mind. There have been many occasions where the hidden sniper has used visible lasers to disperse crowds, or send some angry, poorly armed youths home to their mothers. One of the more common visible laser systems being used in combat today is the M6X combination tactical flashlight and fully boresightable visible laser. Two of the leading manufacturers of this are Insight and Streamlight, and these combination units can be used on pretty much anything from your pistol to your Squad Automatic Weapon (SAW). These tiny lights can be specifically manufactured to mount directly to the frame of your unique sidearm, or when it comes to the rifle, you'll see them attached to the 1913 rail system. Easy on, easy off with the throw of one locking lever, or for more permanent attachment, there are some more fixed options

*Photo courtesty of U.S. Army*

Here a U.S. SF operator is using the SOFLAM to mark a target for airstrike in Afghanistan in 2001.

that might involve an Allen wrench or screwdriver at most.

The SOFLAM AN/PEQ-1 can be used for ranging and target designation. It is equipped with a 10 power scope, mounting rails, and weighs about 12 pounds, depending on what you have attached to it. It can be used with or without a tripod and can mark a target up to 20,000 meters away (you'll want it on the tripod for that distance). It can take a hard bounce without damage and is pretty easy to carry. The main "spot" on a target at 5 kilometers is only about 2.3 meters square, so the precision of any ordnance is within that target square. That's why laser guided bombs come through the window sometimes. The SOFLAM AN/PEQ-1 was originally manufactured by Litton, but like most laser products, is now being manufactured by several leading companies. Either way, it is widely used and well known on the battlefield, and provides politicians with the ability to actually get a human intelligence (HUMINT) regarding the effectiveness of the strike along with actual and collateral damage. The laser inside is a Nd:YAG laser, a Neodymium:Yttrium Alumina Garnet Laser, which is to say that it's a solid laser that uses a laser diode to start the lasing of the main rod of Nd:YAG to create the beam. The beam is fairly efficient and the NiCad battery in the SOFLAM AN/PEQ-1 lasts a reasonable amount of time, and to add in even more cool factor, the unit can be operated remotely, so during a covert operation it could be hidden and left behind to be activated at your leisure from a safe location.

One of the most effective tools of the modern sniper is intimidation; the enemy knows the sniper is out "there" somewhere. It has been proven again and again over the course of history that trained snipers can crush the morale of entire armies. A brightly visible laser spot on a forehead makes everyone around re-evaluate the safety of their current location. Maybe tea with mom is a better place to be?

11

# Gear

What gear you bring with you in the field depends completely on what the mission is. Are you doing sniper overwatch for a direct action mission? Reconnaissance and surveillance, and if so, for how long, and what's the environment? Urban? Desert? There are some things though that will be with you on any sniper mission. To use military speak, your kit is really broken down into three levels called lines of gear.

First line gear is what is attached to your body, from your boots to your belt and whatever you can jam into your pockets. Lets break it out in head to toe fashion. There is one universal expression when it comes to outerwear and additional gear you choose whether or not to bring with you: "Travel light, freeze at night." Sure, it may have been 70°F during the day in Iraq, but it *is* the desert, and if you end up staying the night, you had better be aware that the temp just might get down in the 30s. Your head is the most important part of your body; best to keep it solidly between your shoulders. For you white boys operating in the desert, remember shade is your friend, and it's also nice to be able to have a lid on your hat that will cover the eyepiece of the scope so that the sun won't reflect off it and blind you.

Note the lid of the author's hat just covers the edge of the optical eyepiece to prevent glare.

Floppy hats, baseball hats, ghillie hats for certain operations . . . these all work. When the sun goes down, though, it's best to have a black beanie tucked into your cargo pocket in your first line or at the least in your second or third line gear. You won't regret it. Sunglasses are key, too: Don't skimp on quality. Your vision is depending on it, and as a sniper, keeping your eyesight as sharp as possible and maintaining that 20/10 eyesight you were blessed with is vital. There has been a push by the big military for ballistic-quality lenses in the protective eyewear for the troops. This is a good thing but may not be required or practical for every mission. A lot of mainstream eyewear companies are starting to release great lenses that offer fantastic optical quality in conjunction with ballistic safety. Since we're working head to toe, we would feel remiss if we didn't mention that you might be in a place for a long time, and it's not like you can (or should) fire up a cigarette. Chewing gum, on the other hand, is another story. During the second Gulf War on the push from the Kuwait border toward Baghdad, the authors personally witnessed a teammate chew *one* piece of Trident spearmint gum for 21 straight days. ONE PIECE. That story just needed to be told. Let's drive on with gear.

Digital camouflage patterns have been proven to increase not only the time to detection of an object, but the recognition of the object as well.

Camouflage clothing is constantly changing, as it should. The new Multicam that the majority of the U.S. military is using is a big leap forward in technology with regards to camouflage and how the eye views objects. This, too, is mission dependent, but the new colors blended into the digital camouflage pattern work very well in a wide variety of environments.

Other nice touches are being built into the new uniforms that were adapted from how SpecOps guys were modifying their uniforms 15 years ago. The pockets that are now on both shoulders are great, and you'll also notice that they are canted forward for easy access. One of the problems guys operating in covert environments have to deal with, though, is Velcro. The fact is, if you needed something out of your Velcro pocket but needed to maintain an extremely quiet profile you could be in trouble. Certain guys will swap the velco out for buttons when they know they will repeatedly be heading into environments where noise discipline is paramount. There are two canted pockets on the chest, but the pockets that used to be below those have been removed for operators, as you are usually tucking your cammie top into your pants. In the old days guys would just remove them. A pocket isn't any good when tucked into your crotch, and the extra material is annoying. A good rigger's belt is nice: It's secure and in a pinch can be used to rappel or in a hoist operation. Gloves are always important and essential if you're doing a long stalk through dicey terrain. Lots of guys are using gloves designed for professional baseball or football, but there are also new tactical gloves being sold that offer the best of most worlds. They may have a carbon fiber protective plate across the back of the knuckles, some extra protection on the palm, and well designed fingers for increased sensitivity and dexterity.

Footwear is a personal choice, and there really is no right boot or shoe for sniper ops. Mission dependent. In one leg pocket you should have some sort of medical blowout kit for self aid that includes a tourniquet and some blowout bandages. Somewhere in your first line you should have your bloodchit. A bloodchit is printed on waterproof, tearproof paper, and the four edges are perforated with tear away pieces. These pieces have a serial number imprinted on them that matches the main chit and can

The next generation in protective materials from Neptunic Technologies in San Diego, CA. There are materials in these gloves that withstand fire, frag, and edged weapons . . . pretty cool.

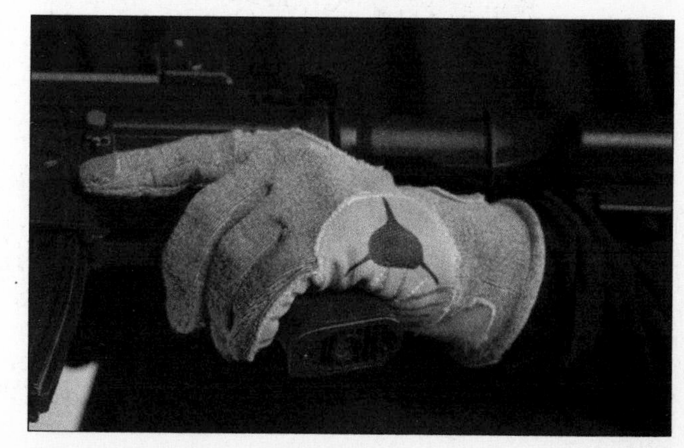

be given to anyone you deem worthy in an emergency who assists you, and can be turned in for a reward at a U.S. government installation. Each chit's ID number is logged and associated with an individual, so that if a corner does get turned in, we will know exactly who is on the run and that they were at least alive recently. In addition to the large American flag on the front of the paper, it also says in many languages relative to the region you are operating in something along the lines of:

> I am a citizen of the United States of America. I do not speak your language. Misfortune forces me to seek your assistance in obtaining food, shelter, and protection. Please take me to someone who will provide for my safety and see that I am returned to my people. My government will reward you.

In another pocket you don't necessarily want to pack your George Costanza wallet, but have your Department of Defense common access card (DOD CAC) identification and a decent amount of cash (in both United States and local currency). If the shit hits the proverbial fan and you are on your escape and evasion (E&E) plan, cash might be the only thing getting you that one item you need to survive. Money talks, and that expression is globally accepted. And if you are on a covert operation, you might not be wearing a uniform, and the ID will be the only thing that gets you back into fortified friendly zones. A small survival kit should take up your other cargo pocket space. Something to eat (even if it's just a Clif Bar), some water purification tablets, a good pocketknife, a signal mirror, some cammie paint, a good lighter . . . you get the idea. It's also nice to have a small handheld GPS and perhaps a map in case you get in trouble . . . you'll want those things if you have to call in an emergency close air support

**An example of an old bloodchit**

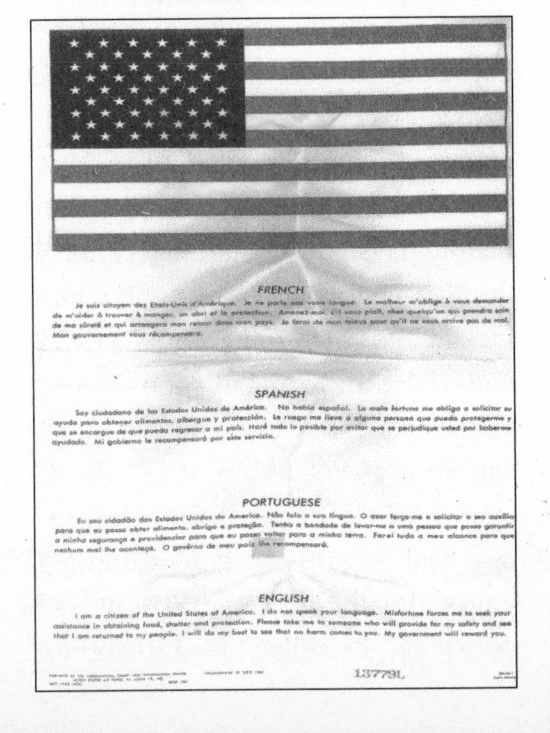

FRENCH

Je suis citoyen des Etats-Unis d'Amérique. Je ne parle pas votre langue. Le malheur m'oblige à vous demander de m'aider à trouver à manger, un abri et la protection. Amenez-moi, s'il vous plaît, chez quelqu'un qui prendra soin de ma sûreté et qui arrangera mon retour dans mon pays. Je ferai de mon mieux pour qu'il ne vous arrive pas de mal. Mon gouvernement vous récompensera.

SPANISH

Soy ciudadano de los Estados Unidos de América. No hablo español. La mala fortuna me obliga a solicitar su ayuda para obtener alimentos, albergue y protección. Le ruego me lleve a alguna persona que pueda protegerme y que se encargue de que pueda regresar a mi país. Haré todo lo posible por evitar que se perjudique usted por haberme ayudado. Mi gobierno le recompensará por éste servicio.

PORTUGUESE

Eu sou cidadão dos Estados Unidos da America. Não falo a sua lingua. O azar força-me a solicitar o seu auxílio para que eu possa obter alimento, abrigo e proteção. Tenha a bondade de levar-me a uma pessoa que possa garantir a minha segurança e providenciar para que eu possa voltar para a minha terra. Farei tudo a meu alcance para que nenhum mal lhe aconteça. O govêrno de meu paiz lhe recompensará.

ENGLISH

I am a citizen of the United States of America. I do not speak your language. Misfortune forces me to seek your assistance in obtaining food, shelter and protection. Please take me to someone who will provide for my safety and see that I am returned to my people. I will do my best to see that no harm comes to you. My government will reward you.

13779L

A chest rig, shown here by Glen, can carry a lot of kit comfortably. Peltor earmuff headset works great for limiting eardrum-shattering noise and accentuating other sounds.

(CAS) mission to save your ass or to take out a significant target.

Your second line gear is really your fighting kit. This will include your primary weapon, and perhaps your secondary as well if you have a pistol slid into your pocket somewhere. Most likely you will have some type of load bearing equipment (LBE), too. A Rhodesian chest rig or similar is popular among snipers, as it keeps your hips free for ease of movement, and you can carry pretty much everything you need for a lot of ops. What a lot of guys would refer to as a slick rig.

This line of kit would also have an IR/overt strobe light for marking your position to friendlies, night vision, extra batteries for optics or night vision goggles, water, extra ammunition, binos, a range finder if not built into your binoculars, and perhaps some 550 cord/bungie cords/zip ties/rubber bands/paperclips/thumbtacks to help set up a shooting position or hide site. Some type of comms will be stashed in your rig, too, whether it's a handheld radio, cell phone, or other device. Communications can definitely be your most powerful weapon. A sniper can feel well endowed when in a strong shooting position with a .50 cal rifle loaded and ready, but controlling a flight of F-18s for CAS strikes is truly powerful and offers the potential for quite a difference in firepower.

Third line gear is equipment that you might carry in a backpack or drag bag. The variety of material to be included in this could fill another book and is dependent on what you've been tasked with.

This is another option for a "chest rig." Tactical Assault Gear (TAG) is a San Diego based company owned by Chris Osman that makes great product. This is their Split Front Chest Rig.

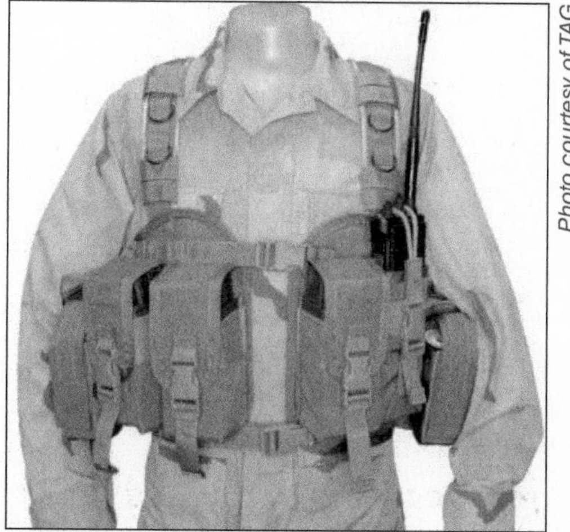

*Photo courtesy of TAG*

A digital SLR camera with a telephoto lens might be part of the kit, with a ruggedized laptop and a satellite uplink for sending photos or video anywhere in the world. The powers that be enjoy being kept in the loop when a small team is in the field, and with that equipment you can provide detailed imagery customized for your audience. A sniper in a forward observation position could identify and photograph primary breach points on a target, patrol patterns, enemy numbers, equipment, and uniforms. This is pertinent and powerful information for follow-on missions. If you and your team know that you are going to be in the field for an extended stay, your third line can and will grow to absurd sizes. It's not unusual to see someone humping in a one-hundred-pound ruck or more. Hide site material, digging equipment, enough batteries to power all the electronic equipment, antennae, food, and the most important and heaviest of them all: water. The most prominent conflicts the United States is currently engaged in are in desert climates. Water is critical, especially for the shooter. The first thing that starts degrading from a physiological standpoint when dehydration starts to set is your vision. Stay hydrated and keep that clear sight picture.

This dragbag by RS Tactical can be worn as a backpack when your sniper rifle cannot (and should not) be used as your primary weapon for long infils or for close quarters until in your final shooting position.

Photo courtesy of Medford Knife & Tool

Every sniper needs a good backup blade for close up action or cracking open a cold beer after a long op. This is the Tactical Tanto-1 by Medford Knife & Tool (www.medfordknife.com). Coauthor Brandon Webb is currently working on a signature blade with Medford.

12

# The Morality of Killing

Too often we get asked what it's like to kill someone or take another human being's life. This is a very difficult and uncomfortable question to answer and we encourage you *not* to ask it to anyone unless you share a close relationship with that person. We did think it was important enough to at least address the topic of killing and give you our perspective on the issue.

What are morals? They involve culturally accepted standards of what is right and what is wrong. The only thing that is certain is that there exists no set of universally accepted morals. In fact they vary widely from culture to culture. In some countries what is right is wrong in others. For instance, it is still acceptable for a man to kill his wife in some Middle Eastern countries. So how do we answer this question?

We have a common belief in the Special Operations community that there are three types of people in the world: wolves, sheepdogs, and sheep. The wolves are what most people would consider evil people: They are the rapists and murderers of the world. They prey on the weak, and they use violence and fear to achieve their goals. Then there are the sheepdogs. They look similar to the wolf and can easily be mistaken for them, but they are there to protect those that cannot protect themselves. They exist solely to look after the flock. The sheep are everyday good people who go about their lives in safety because they are protected. They don't necessarily like to be around or acknowledge the sheepdogs because this makes them nervous and aware that the wolves are out there ready to cause harm. The Special Operations personnel of the world are definitely sheepdogs. As Winston Churchill said, "We sleep soundly in our beds because rough men stand ready in the night to visit violence on those who would do us harm."

From our personal experience and extensive conversations with other operators and snipers, everyone deals with killing and death differently. Much of this will be a reflection of the unique individual, their upbringing, and the situation that brought them to the killing fields. Whereas the snipers of Stalingrad were fighting for their very lives and for that of their homeland and families, the German snipers they faced had decidedly different motivations. Major Hesketh Pritchard, the World War I–era British sniper, sums it up well in Andy Dougan's *Through the Crosshairs*:

> The sniper must be able to calmly and deliberately kill targets that may not pose an immediate threat to him. It is much easier to kill in self-defense or in the defense of others than it is to kill without apparent provocation. The sniper must not be susceptible to emotions such as anxiety or remorse. Candidates whose motivation towards sniper training rests

No matter your beliefs, the core values of the major religions remain the same. Despite the benevolence of these basic tenets, killing in the name of a God is a historical and modern fact.

> mainly in the desire for prestige may not be capable of the cold rationality the sniper's job requires.

During SEAL sniper training, you learn a system and become incredibly proficient at the mental and mechanical side of making a long-range shot at a silhouette target. There were several "celebrity" coaches and trainers that cycled through the three-and-half month course, and it was one of these men that was really the only one who ever brought up and discussed at length the mental aspect and potential repercussions of taking someone's life. He was a decorated combat veteran, with over fifty confirmed kills during multiple deployments in Vietnam and later became a champion long-range competition shooter. He was also a religious man, and said:

> When I was over there, in the bush, watching and waiting for my next target, you had a fair amount of time to reflect on your life, and that of the man whose life you were going to attempt to take. I figured I was trained to do a job, just like my VC counterpart, and if and when that person took shape in my crosshairs, well then it was just his time, and God was calling him home. I also realized that on any given day it could be my turn in someone else's crosshairs, and I held no ill will towards the man that held that rifle, as he was just doing his job too, and it was just my time to go home.

It was a powerful message delivered by a humble man, and it resonated with everyone in the class, from the atheist to the practicing Christian. The one thing we can tell you is that 99 percent of the people we know who are Special Operations snipers have no problem rationalizing the kill. We have met some guys who have a problem with it and these guys ultimately get transitioned out of the Special Operations community into roles they are more suited for. Most are quickly forgotten and never heard from again. In our community, the training is long and intense, and someone would have to be pretty dense and obtuse to not reflect on what they were being trained to do. Those people usually don't make it to our community, and those that do have already rationalized their job in their minds. However, thinking about it during training and experiencing your first

Photo courtesy of U.S. Navy

**The USS Cole is an awful reminder to remain vigilant. Bad people are out there and are full of hate toward those that believe differently than they do.**

**The cemetery at Colleville-sur-Mer in Normandy, France, on a bluff overlooking Omaha Beach contains 9,387 American military dead.**

combat are different beasts indeed, and there will be a moment when, despite all the training you went through, you will take a collective breath and be stripped down to the most simple truths: You are a man, recognizing another man who is now dead because of your actions. This is the moment of truth and the time where individual rationalizations will take hold, and your next sleep will either be childlike or completely disturbed. In Sasser and Roberts's *One Shot One Kill*, a Marine captain and legendary Vietnam–era sniper and instructor reflects on the type of mindset a man must have to be a great sniper:

> There is no hate of the enemy. Psychologically, the only motive that will sustain the sniper is knowing he is doing a necessary job and having the confidence that he is the best person to be doing it. When you look through a scope the first thing you see is the eyes. There is a lot of difference between shooting at an outline . . . and shooting at a pair of eyes. Many men can't do it at that point.

As recounted in Dougan's *Through the Crosshairs*, Carlos Hathcock sums up his thoughts about being a sniper in his characteristically simple and still-relevant way:

> I like shooting, and I love hunting. But I never did enjoy killing anybody. It's my job. If I don't get those bastards, then they are gonna kill a lot of those kids dressed up like Marines. That's the way I look at it . . . But I never went on any mission with anything in mind other than winning this war and keeping those . . . bastards from killing more Americans.

13

# The Future of Special Operations: Sniping into the 21st Century

It's amazing and scary that the U.S. military and state and federal law enforcement communities do not have standardized sniper programs in place. I'm not saying that there aren't great programs out there. The U.S. Navy SEAL sniper program is one of the best in the world. What I'm pointing out is that the U.S. sniper community needs to get it together and standardize training and methodology. For the most part, the military is close in this regard but I'm shocked at the lack of standardization and low training standards in the American law enforcement communities. I've worked with some of the biggest and well-funded law enforcement SWAT sniper programs and have been shocked at the lack of training these solid operators receive. I have yet to run across a law enforcement sniper program that has incorporated the best practices of the sniper community.

Military snipers have the luxury of the battlefield environment and favorable rules of engagement; in most cases we aren't concerned with hostages and other casualties involved in the shots we take. My SEAL friend Jason (a natural born killer) just returned from Afghanistan where his unit killed close to 200

This is a great example of modern advances in nanotech camouflage technology. You can see the person's camo jacket blending in with the surrounding environment.

bad guys. On one OP he said he got two guys with one shot. He shot the driver and killed the passenger who was inline and shot with the exiting round (we call this bullet path).

Fortunately the passenger up front needed killing too. I think you get the point I'm trying to make here and it is that law enforcement has different rules of engagement and usually the stakes are much higher because citizen hostages are involved. This is a critical situation where there is zero room for error. Most of you will remember the recent Somali pirates that were shot to rescue US shipping Capt. Phillips. Three SEAL snipers shot the pirates in total darkness from one moving platform to another moving platform, with flawless coordinated surgical precision. Three simultaneous kill shots that ultimately diffused an international situation and led to Phillips' rescue. These snipers were extremely well trained and they accounted for every factor both human and environmental, guaranteeing success.

I find that most of the local law enforcement agencies let ego and rivalries get in the way of solid training and in my experience most local law enforcement sniper programs are severely lacking. I've been on the range with qualified snipers that do not even understand the fundamentals of internal, external, or terminal ballistics. Most argue with me for wanting to push their training past the 100-meter threshold. I recognize that the average police sniper shot is around 80 yards, but it is also extremely important to train beyond practical application. If you can consistently drive nails at 600 yards under adverse conditions (artificially induced stress, wind, barometric pressure, high angle, etc.), the shot you make when the call comes will be like walking the dog. I am baffled when people try and argue with me about the reasoning behind this methodology. If you look at every world-class athletic program, they all train much harder then what is required in competition.

It is my hope that the U.S. law enforcement sniper community realizes the importance of standardizing their training and how important the role of the sniper is when the time comes.

One thing that is for sure is that the sniper and the role he or she plays is here to stay. A stealthy and precise kill shot to eliminate a threat is what the job is all about. The target will be dead before they hit the ground. As the threat of terrorism increases, I see an increase in employment into the 21st century in both the military and law enforcement communities. It's not that we live in a more dangerous world, but rather that as we have become more of a global society, the threat to society is much more asymmetric in nature. In fact, globalization in my opinion is threatening more traditional societies that in turn look to combat the threat of modern society with terrorism. It is in most cases the only way to effectively engage more affluent societies. As this trend increases affluent societies will have to adapt in an attempt to deal with the modern asymmetric threat we face. A part of this adaptation is the increased focus and use of unconventional strategy and tactics. A definite tool in the unconventional tool box is the sniper. While the weapons and technology employed by the modern-day sniper are rapidly changing, the mission is still the same: delivering precise solutions with zero collateral damage and maximum effectiveness. The modern-day sniper is here to stay.

Radical changes are occurring in technology that can be leveraged by the sniper community. We now have technology in optics that fuses together the IR and thermal spectrum, nanotech camouflage that bends light, and lasers coupled to sniper scopes that are capable of precision wind calculations. It's exciting to see this new technology and how it can increase the modern-day sniper's capacity and effectiveness on the job.

The problem has never been innovation; it has been getting the most current technology in the hands of the operators. It is unfortunate that the bureaucracy involved with govern-

Photo courtesy of Ken Loving

**Friend of the authors Ken Lovings' 300 Remington Ultra Mag. This tricked-out weapon system could easily reach out well past 1,000 yards and is seen here scoping out the New Mexico high country.**

ment acquisition often gets in the way. Typically large systems integrators for the government deliver untimely and irrelevant solutions. I can remember advising a large billion-dollar defense company that was working on a computer system for snipers. What they had been developing for years with tens of millions of U.S. taxpayer dollars I could get off the shelf for thousands. They were so focused on their program that they were blind to this fact, and even after coming to terms with this, the program director said he was going forward anyway because without the program his job would go away! Complete madness and a tremendous waste of taxpayer dollars. The point I'm trying to make is that the U.S. government acquisition system is broken and we need to fix it or else our enemies will be better equipped than us. I have personally been involved on the military side and fortunately Admiral Olsen of U.S. Special Operations Command (U.S. SOCOM) has recognized that this is a problem and SOCOM is in the process of fixing it as you read this book.

APPENDIX

A

# If You Can't Afford a Ballistic Computer, or Even if You Can

In this appendix are sniper data charts that were given to us by Dave Durham at CheyTac. Knowing what your rifle does with the ammunition you choose to shoot, or that you are issued, as well as how it performs in different conditions is paramount to making accurate shots, especially if you adhere to the "one shot, one kill" rule.

Keeping accurate logs of your shooting will be a big help in knowing how you and your rifle will perform under a variety of conditions. Most shooters will call these "dope" books. These will start falling by the wayside as ballistic software becomes more prevalent in the world, but let's not forget how important it is to embrace the old before learning the new. Batteries die, electromagnetic pulses fry electronics . . . so keep your round count and dope books up to date.

# Sniper Chart One: Barrel Log

**Barrel Log**

| Date | Ammunition | Lot | # Fired | Total | Date | Ammunition | Lot | # Fired | Total |
|---|---|---|---|---|---|---|---|---|---|
| | | | | | | | | | |
| | | | | | | | | | |
| | | | | | | | | | |
| | | | | | | | | | |
| | | | | | | | | | |
| | | | | | | | | | |
| | | | | | | | | | |
| | | | | | | | | | |
| | | | | | | | | | |
| | | | | | | | | | |
| | | | | | | | | | |
| | | | | | | | | | |

| Notes: | | |
|---|---|---|
| | This Page | |
| | Previous Page | |
| | Total | |

**Barrel Log**

| Date | Ammunition | Lot | # Fired | Total | Date | Ammunition | Lot | # Fired | Total |
|---|---|---|---|---|---|---|---|---|---|
| | | | | | | | | | |
| | | | | | | | | | |
| | | | | | | | | | |
| | | | | | | | | | |
| | | | | | | | | | |
| | | | | | | | | | |
| | | | | | | | | | |
| | | | | | | | | | |
| | | | | | | | | | |
| | | | | | | | | | |
| | | | | | | | | | |
| | | | | | | | | | |

| Notes: | | |
|---|---|---|
| | This Page | |
| | Previous Page | |
| | Total | |

## Sniper Chart Two: Come Ups

| Come-Up | From 0 | Come-Up | From 0 | 10 MPH |
|---|---|---|---|---|
| 100 | | 150 | | |
| 200 | | 250 | | |
| 300 | | 350 | | |
| 400 | | 450 | | |
| 500 | | 550 | | |
| 600 | | 650 | | |
| 700 | | 750 | | |
| 800 | | 850 | | |
| 900 | | 950 | | |
| 1000 | | 1050 | | |

**TEMP RANGE:**

__________ deg F

**LOAD DATA:**

Bullet: __________
Velocity: __________
Alttitude: __________

**NOTES:**

| Come-Up | From 0 | Come-Up | From 0 | 10 MPH |
|---|---|---|---|---|
| 100 | | 150 | | |
| 200 | | 250 | | |
| 300 | | 350 | | |
| 400 | | 450 | | |
| 500 | | 550 | | |
| 600 | | 650 | | |
| 700 | | 750 | | |
| 800 | | 850 | | |
| 900 | | 950 | | |
| 1000 | | 1050 | | |

**TEMP RANGE:**

__________ deg F

**LOAD DATA:**

Bullet: __________
Velocity: __________
Alttitude: __________

**NOTES:**

# Sniper Chart 3

Formulas

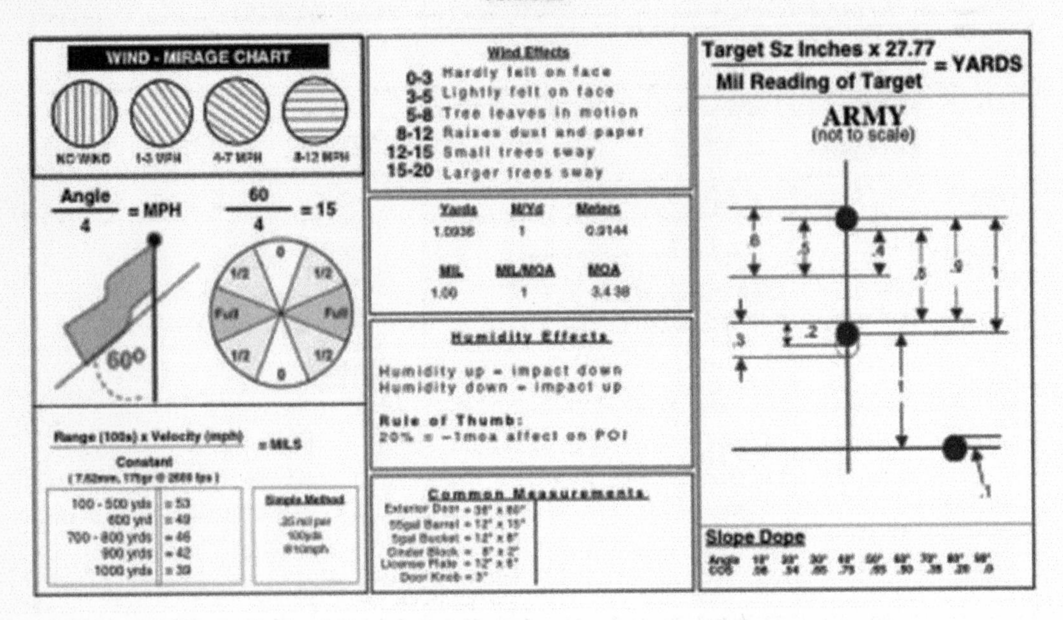


Formulas

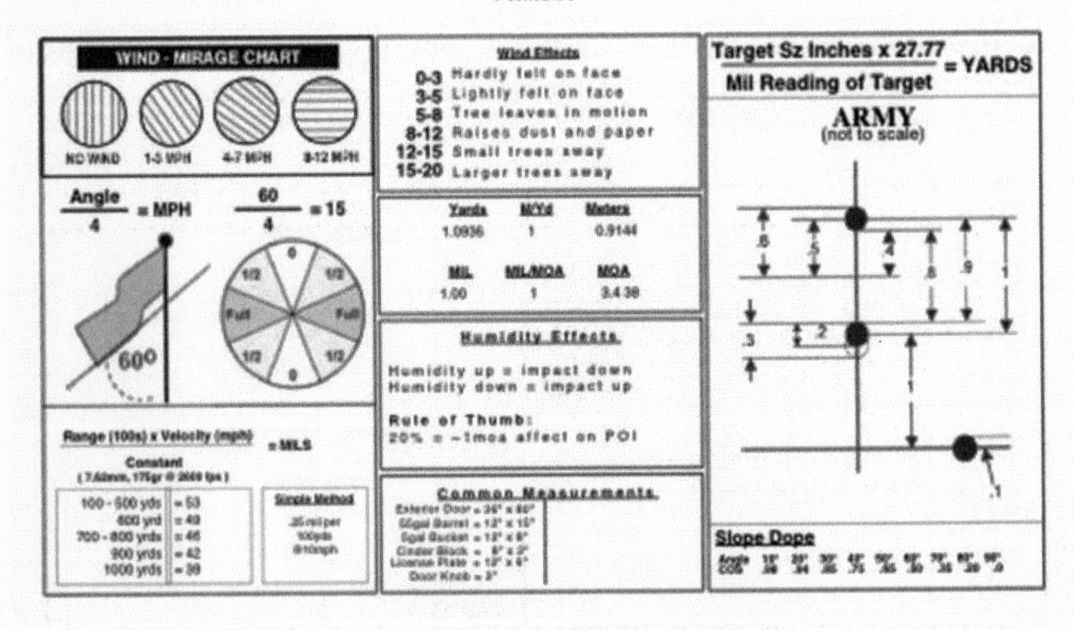

## Sniper Chart 4: Zero Data

**Zero Data**

| Date | Time | Location | Rifle/Scope | Ammunition | Distance |
|---|---|---|---|---|---|
| | | | | | |

| Altitude | Humidity | Baro. Press. | Temp | Mirage | Light | Wind |
|---|---|---|---|---|---|---|
| | | | | | ☐ Bright<br>☐ Hazy<br>☐ Overcast<br>☐ Changing | ☐ Light (3)<br>☐ Med (7)<br>☐ Heavy (15)<br>☐ ____ |

| Shot # | 1 | 2 | 3 | 4 | 5 | 6 | 7 | 8 | 9 | 10 |
|---|---|---|---|---|---|---|---|---|---|---|
| Elev. | | | | | | | | | | |
| Wind | | | | | | | | | | |
| Call | ☐ | ☐ | ☐ | ☐ | ☐ | ☐ | ☐ | ☐ | ☐ | ☐ |
| **Shot #** | **11** | **12** | **13** | **14** | **15** | **16** | **17** | **18** | **19** | **20** |
| Elev. | | | | | | | | | | |
| Wind | | | | | | | | | | |
| Call | ☐ | ☐ | ☐ | ☐ | ☐ | ☐ | ☐ | ☐ | ☐ | ☐ |

Notes:

Target Size =

**Zero Data**

| Date | Time | Location | Rifle/Scope | Ammunition | Distance |
|---|---|---|---|---|---|
| | | | | | |

| Altitude | Humidity | Baro. Press. | Temp | Mirage | Light | Wind |
|---|---|---|---|---|---|---|
| | | | | | ☐ Bright<br>☐ Hazy<br>☐ Overcast<br>☐ Changing | ☐ Light (3)<br>☐ Med (7)<br>☐ Heavy (15)<br>☐ ____ |

| Shot # | 1 | 2 | 3 | 4 | 5 | 6 | 7 | 8 | 9 | 10 |
|---|---|---|---|---|---|---|---|---|---|---|
| Elev. | | | | | | | | | | |
| Wind | | | | | | | | | | |
| Call | ☐ | ☐ | ☐ | ☐ | ☐ | ☐ | ☐ | ☐ | ☐ | ☐ |
| **Shot #** | **11** | **12** | **13** | **14** | **15** | **16** | **17** | **18** | **19** | **20** |
| Elev. | | | | | | | | | | |
| Wind | | | | | | | | | | |
| Call | ☐ | ☐ | ☐ | ☐ | ☐ | ☐ | ☐ | ☐ | ☐ | ☐ |

Notes:

Target Size =

# Sniper Chart 5: Zero Summary

Zero Summary Chart

☐ Yards ☐ Meters — Temperature (F) — Ammunition:

| Distance | 20° | 30° | 40° | 50° | 60° | 70° | 80° | 85° | 90° | 100° | 105° |
|---|---|---|---|---|---|---|---|---|---|---|---|
| 50 | | | | | | | | | | | |
| 75 | | | | | | | | | | | |
| 100 | | | | | | | | | | | |
| 125 | | | | | | | | | | | |
| 150 | | | | | | | | | | | |
| 175 | | | | | | | | | | | |
| 200 | | | | | | | | | | | |
| 225 | | | | | | | | | | | |
| 250 | | | | | | | | | | | |
| 275 | | | | | | | | | | | |
| 300 | | | | | | | | | | | |
| 325 | | | | | | | | | | | |
| 350 | | | | | | | | | | | |
| 375 | | | | | | | | | | | |
| 400 | | | | | | | | | | | |
| 425 | | | | | | | | | | | |
| 450 | | | | | | | | | | | |
| 475 | | | | | | | | | | | |
| 500 | | | | | | | | | | | |

Zero Summary Chart

☐ Yards ☐ Meters — Temperature (F) — Ammunition:

| Distance | 20° | 30° | 40° | 50° | 60° | 70° | 80° | 85° | 90° | 100° | 105° |
|---|---|---|---|---|---|---|---|---|---|---|---|
| 50 | | | | | | | | | | | |
| 75 | | | | | | | | | | | |
| 100 | | | | | | | | | | | |
| 125 | | | | | | | | | | | |
| 150 | | | | | | | | | | | |
| 175 | | | | | | | | | | | |
| 200 | | | | | | | | | | | |
| 225 | | | | | | | | | | | |
| 250 | | | | | | | | | | | |
| 275 | | | | | | | | | | | |
| 300 | | | | | | | | | | | |
| 325 | | | | | | | | | | | |
| 350 | | | | | | | | | | | |
| 375 | | | | | | | | | | | |
| 400 | | | | | | | | | | | |
| 425 | | | | | | | | | | | |
| 450 | | | | | | | | | | | |
| 475 | | | | | | | | | | | |
| 500 | | | | | | | | | | | |

Zero Summary Chart

☐ Yards ☐ Meters

Temperature (F)

Ammunition:

| Distance | 20° | 30° | 40° | 50° | 60° | 70° | 80° | 85° | 90° | 100° | 105° |
|---|---|---|---|---|---|---|---|---|---|---|---|
| 50 | | | | | | | | | | | |
| 75 | | | | | | | | | | | |
| 100 | | | | | | | | | | | |
| 125 | | | | | | | | | | | |
| 150 | | | | | | | | | | | |
| 175 | | | | | | | | | | | |
| 200 | | | | | | | | | | | |
| 225 | | | | | | | | | | | |
| 250 | | | | | | | | | | | |
| 275 | | | | | | | | | | | |
| 300 | | | | | | | | | | | |
| 325 | | | | | | | | | | | |
| 350 | | | | | | | | | | | |
| 375 | | | | | | | | | | | |
| 400 | | | | | | | | | | | |
| 425 | | | | | | | | | | | |
| 450 | | | | | | | | | | | |
| 475 | | | | | | | | | | | |
| 500 | | | | | | | | | | | |

Zero Summary Chart

☐ Yards ☐ Meters

Temperature (F)

Ammunition:

| Distance | 20° | 30° | 40° | 50° | 60° | 70° | 80° | 85° | 90° | 100° | 105° |
|---|---|---|---|---|---|---|---|---|---|---|---|
| 50 | | | | | | | | | | | |
| 75 | | | | | | | | | | | |
| 100 | | | | | | | | | | | |
| 125 | | | | | | | | | | | |
| 150 | | | | | | | | | | | |
| 175 | | | | | | | | | | | |
| 200 | | | | | | | | | | | |
| 225 | | | | | | | | | | | |
| 250 | | | | | | | | | | | |
| 275 | | | | | | | | | | | |
| 300 | | | | | | | | | | | |
| 325 | | | | | | | | | | | |
| 350 | | | | | | | | | | | |
| 375 | | | | | | | | | | | |
| 400 | | | | | | | | | | | |
| 425 | | | | | | | | | | | |
| 450 | | | | | | | | | | | |
| 475 | | | | | | | | | | | |
| 500 | | | | | | | | | | | |

APPENDIX

B

# Velocity Comparison Tables

"I know what you're thinking, punk. You're thinking, 'Did he fire six shots or only five?' Now to tell you the truth I forgot myself in all this excitement. But being this is a .44 Magnum, the most powerful handgun in the world and will blow you head clean off, you've gotta ask yourself a question: 'Do I feel lucky?' Well, do ya, punk?"

—*Dirty Harry*

That movie was a long time ago and today there are even more powerful handgun cartridges available for the big bore pistol shooters. But what does a big, loud, high recoil have do to with snipers? Nothing! However, if we look at the ballistic table again, we can see just how powerful today's sniper rifles are. Table 1 compares the velocities and energies of the .44 Mag rifle cartridge to sniper cartridges. Table 2 compares the time to target and the drop from the bore of the .44 Mag to the same rounds.

## Table 1: Velocity and Energy Comparisons

| Cartridge BC | Bullet Wt | MV | ME | V @100 | E @100 | V @300 | E @300 | V @ 500 | E @500 | V @1000 | E @1000 |
|---|---|---|---|---|---|---|---|---|---|---|---|
| 44 Mag/0.157 | 240 | 1,760 | 1,650 | 1,361 | 986 | 951 | 482 | 784 | 328 | 528 | 148 |
| 223 Rem/0.372 | 77 | 2,750 | 1,293 | 2,503 | 1,071 | 2,049 | 717 | 1,645 | 463 | 1,007 | 173 |
| 308/0.458 | 168 | 2,600 | 2,521 | 2,410 | 2,166 | 2,055 | 1,574 | 1,733 | 1,120 | 1,142 | 486 |
| 308/0.470 | 167 | 2,690 | 2,683 | 2,510 | 2,320 | 2,146 | 1,708 | 1,824 | 1,233 | 1,201 | 535 |
| 300WM/0.533 | 190 | 2,950 | 3,671 | 2,773 | 3,243 | 2,439 | 2,509 | 2,129 | 1,912 | 1,460 | 899 |
| 338/0.587 | 250 | 2,950 | 4,830 | 2,789 | 4,316 | 2,484 | 3,424 | 2,198 | 2,682 | 1,532 | 1,303 |
| 338/0.675 | 250 | 2,969 | 4,892 | 2,828 | 4,438 | 2,559 | 3,634 | 2,305 | 2,949 | 1,739 | 1,678 |
| 338/0.768 | 300 | 2,800 | 5,222 | 2,680 | 4,785 | 2,451 | 4,001 | 2,232 | 3,318 | 1,734 | 2,002 |
| 408 Chey Tac/1.0 | 419 | | 7,700 | | | | | | | | |
| 50 BMG/1.05 | 750 | | 11,200 | | | | | | | | |

## Table 2: Time to Target and Drop Comparison

(Note the drop for all distance is negative!)

| Cartridge | BC | Bullet weight | Time to 100 | Drop @ 100 | Time to 500 | Drop @ 500 | Time to 2000 | Drop @ 2000 |
|---|---|---|---|---|---|---|---|---|
| .44 Mag | 0.157 | 240 | 0.1942 s | 6.6 inches | 1.4359 s | 319.4 inches | 12.99 s | 20,496.6i inches |
| .223 Rem | 0.372 | 77 M. King | 0.1142 s | 2.4 inches | 0.7058 s | 81.6 inches | 5.69 s | 4,480.8 inches |
| .308 Win | 0.458 | 168 M. King | 0.1198 s | 2.7 inches | 0.7073 s | 84.6 inches | 5.09 s | 3,663.8 inches |
| .308 Win | 0.47 | 167 Scenar | 0.1156 s | 2.5 inches | 0.6777 s | 78.1 inches | 4.90 s | 3,363 inches |
| .300 Win Mag | 0.533 | 190 M. King | 0.1049 s | 2.1 inches | 0.5988 s | 62.2 inches | 4.26 s | 2,483 inches |
| .338 Lapua | 0.587 | 250 M. King | 0.1046 s | 2.1 inches | 0.5893 s | 60.8 inches | 4.05 s | 2,229.6 inches |
| .338 Lapua | 0.675 | 250 Scenar | 0.1035 s | 2.0 inches | 0.5735 s | 58.4 inches | 3.64 s | 1,857.4 inches |
| .338 Lapua | 0.768 | 300 Scenar | 0.1095 s | 2.3 inches | 0.6000 s | 64.4 inches | 3.62 s | 1,887.1 inches |

## Table 1

You can see that Harry's .44 Magnum develops 1,650 pound-feet at the muzzle. The .308 at 500 yards has almost as much energy and at 1,000 yards still has one-third the energy of the .44. The .338 Lapua 250-grain Scenar and .300-grain Match King unbelievably have more energy at 1,000 yards than the .44 Mag at the muzzle. Even the hottest hand load for the .44 Mag can only generate a muzzle velocity of approximately 1,400 fps.

## Table 2

In this table you can see that the .44 Mag is not even close to the sniper cartridges in time to target: 13 seconds to reach 2,000 yards and the drop is dramatic—a negative 20,496 inches at 2,000 yards. For you math freaks that's 1,705 feet, or 568 yards. One MOA at 2,000 yards is 20 inches, so the drop would be 20,496 divided by 20—equally 1,024 MOA of elevation. Can you say "impossible hold over?" Sniper rifles and their cartridges are simply awesome examples of man's ability to find a better way to fight his enemies. Feeling lucky, Harry?

APPENDIX C

# Exbal Ballistics Charts

Looking at the following ballistic table, the 77-grain .223 Remington goes subsonic between 800 and 850 yards for the atmospheric conditions used for this example. The energy at 800 yards is 236 food-pounds and the bullet has dropped 229.0 inches from the bore. One MOA at 800 yards is 8 inches. If the rifle is zeroed at 100 yards, you will need 29 MOA in elevation to be on target at 800 (229 ÷ 8 = 28.625).

Exbal Ballistic Calculator

Black-Hills New Mfg: .223 Remington : Sierra MatchKing 0.224" 77gr HPBT

| | SIGHT-IN DATA | FIELD DATA | | |
|---|---|---|---|---|
| Muzzle Velocity (fps) | 2750 | 2750 | | |
| Bullet Weight (grains) | 77 | | | |
| Sight Height (in) | 1.5 | | | |
| Sight-in Distance (yd) | 100 | | | |
| Altitude (ft) | 0 | 0 | | |
| Temperature (deg F) | 59 | 59 | | |
| Pressure@Sea Level(in Hg) | 29.53 | 29.53 | | |
| Relative Humidity (pct) | 78.0 | 78.0 | | |
| Wind Velocity (mph) | | 0 | | |
| Wind Angle (degrees) | | 180 ( 6.0 O'Clock) | | |
| Incline Angle (degrees) | | 0 | | |
| Moving Target Speed (mph) | | 0.0 | | |
| Ballistic Coefficient | 0.372 | 0.362 | 0.362 | 0.343 |
| Lower Velocity Limit (fps) | 3000 | 2500 | 1700 | 0 |

POINT BLANK RANGE DATA

| TARGET HEIGHT (in) | SIGHT-IN DISTANCE (yd) | POINT BLANK RANGE HIGH (yd) | POINT BLANK RANGE LOW (yd) |
|---|---|---|---|
| 2 | 158 | 96 | 178 |
| 4 | 198 | 114 | 227 |
| 6 | 228 | 127 | 264 |
| 8 | 254 | 139 | 297 |
| 10 | 276 | 153 | 321 |

MAX APPARENT TRAJECTORY (in) 0.1

| TARGET DIST (yd) | SIGHT ADJUSTMENTS ELEV MOA | SIGHT ADJUSTMENTS WIND MOA | SIGHT ADJUSTMENTS LEAD MOA | TRAJECTORY VALUES ELEV (in) | TRAJECTORY VALUES WIND (in) | TRAJECTORY VALUES LEAD (in) | VELOCITY (fps) | ENERGY (ft-lb) | Drop from bore line (in) | ARRIVAL TIME (sec) |
|---|---|---|---|---|---|---|---|---|---|---|
| 0 | 0.00 | 0.00 | 0.00 | -1.5 | 0.0 | 0.0 | 2750 | 1293 | 0.0 | 0.0000 |
| 50 | 0.25 | 0.00 | 0.00 | -0.1 | 0.0 | 0.0 | 2625 | 1178 | -0.6 | 0.0558 |
| 100 | 0.00 | 0.00 | 0.00 | -0.0 | 0.0 | 0.0 | 2503 | 1071 | -2.4 | 0.1143 |
| 150 | 0.75 | 0.00 | 0.00 | -1.3 | 0.0 | 0.0 | 2384 | 972 | -5.6 | 0.1757 |
| 200 | 2.00 | 0.00 | 0.00 | -4.1 | 0.0 | 0.0 | 2269 | 880 | -10.4 | 0.2401 |
| 250 | 3.25 | 0.00 | 0.00 | -8.5 | 0.0 | 0.0 | 2157 | 795 | -16.8 | 0.3079 |
| 300 | 4.75 | 0.00 | 0.00 | -14.9 | 0.0 | 0.0 | 2049 | 717 | -25.2 | 0.3792 |
| 350 | 6.25 | 0.00 | 0.00 | -23.3 | 0.0 | 0.0 | 1943 | 646 | -35.5 | 0.4544 |
| 400 | 8.00 | 0.00 | 0.00 | -34.0 | 0.0 | 0.0 | 1842 | 580 | -48.2 | 0.5336 |
| 450 | 10.00 | 0.00 | 0.00 | -47.3 | 0.0 | 0.0 | 1743 | 519 | -63.5 | 0.6173 |
| 500 | 12.00 | 0.00 | 0.00 | -63.5 | 0.0 | 0.0 | 1645 | 463 | -81.6 | 0.7058 |
| 550 | 14.50 | 0.00 | 0.00 | -82.8 | 0.0 | 0.0 | 1551 | 411 | -102.9 | 0.7997 |
| 600 | 16.75 | 0.00 | 0.00 | -105.8 | 0.0 | 0.0 | 1462 | 366 | -127.8 | 0.8992 |
| 650 | 19.50 | 0.00 | 0.00 | -132.8 | 0.0 | 0.0 | 1379 | 325 | -156.8 | 1.0048 |
| 700 | 22.50 | 0.00 | 0.00 | -164.4 | 0.0 | 0.0 | 1303 | 290 | -190.3 | 1.1167 |
| 750 | 25.50 | 0.00 | 0.00 | -201.1 | 0.0 | 0.0 | 1235 | 261 | -229.0 | 1.2350 |
| 800 | 29.00 | 0.00 | 0.00 | -243.5 | 0.0 | 0.0 | 1174 | 236 | -273.4 | 1.3596 |
| 850 | 32.75 | 0.00 | 0.00 | -292.2 | 0.0 | 0.0 | 1122 | 215 | -324.0 | 1.4903 |
| 900 | 37.00 | 0.00 | 0.00 | -347.8 | 0.0 | 0.0 | 1079 | 199 | -381.6 | 1.6267 |
| 950 | 41.25 | 0.00 | 0.00 | -410.9 | 0.0 | 0.0 | 1041 | 185 | -446.6 | 1.7683 |
| 1000 | 46.00 | 0.00 | 0.00 | -481.9 | 0.0 | 0.0 | 1007 | 173 | -519.7 | 1.9150 |
| 1050 | 51.00 | 0.00 | 0.00 | -561.6 | 0.0 | 0.0 | 978 | 163 | -601.3 | 2.0662 |
| 1100 | 56.50 | 0.00 | 0.00 | -650.4 | 0.0 | 0.0 | 951 | 155 | -692.0 | 2.2219 |
| 1150 | 62.25 | 0.00 | 0.00 | -748.7 | 0.0 | 0.0 | 927 | 147 | -792.3 | 2.3818 |
| 1200 | 68.25 | 0.00 | 0.00 | -857.2 | 0.0 | 0.0 | 906 | 140 | -902.8 | 2.5458 |
| 1250 | 74.50 | 0.00 | 0.00 | -976.4 | 0.0 | 0.0 | 886 | 134 | -1023.9 | 2.7136 |
| 1300 | 81.25 | 0.00 | 0.00 | -1106.6 | 0.0 | 0.0 | 867 | 128 | -1156.1 | 2.8852 |
| 1350 | 88.25 | 0.00 | 0.00 | -1248.5 | 0.0 | 0.0 | 849 | 123 | -1299.9 | 3.0606 |
| 1400 | 95.75 | 0.00 | 0.00 | -1402.5 | 0.0 | 0.0 | 831 | 118 | -1455.9 | 3.2398 |
| 1450 | 103.25 | 0.00 | 0.00 | -1569.1 | 0.0 | 0.0 | 815 | 113 | -1624.5 | 3.4229 |
| 1500 | 111.25 | 0.00 | 0.00 | -1749.0 | 0.0 | 0.0 | 798 | 109 | -1806.4 | 3.6098 |
| 1550 | 119.75 | 0.00 | 0.00 | -1942.7 | 0.0 | 0.0 | 783 | 105 | -2002.0 | 3.8005 |
| 1600 | 128.25 | 0.00 | 0.00 | -2150.7 | 0.0 | 0.0 | 768 | 101 | -2211.9 | 3.9952 |
| 1650 | 137.25 | 0.00 | 0.00 | -2373.6 | 0.0 | 0.0 | 754 | 97 | -2436.8 | 4.1937 |
| 1700 | 146.75 | 0.00 | 0.00 | -2612.0 | 0.0 | 0.0 | 741 | 94 | -2677.2 | 4.3961 |
| 1750 | 156.50 | 0.00 | 0.00 | -2866.6 | 0.0 | 0.0 | 728 | 90 | -2933.7 | 4.6025 |
| 1800 | 166.50 | 0.00 | 0.00 | -3137.9 | 0.0 | 0.0 | 715 | 87 | -3207.0 | 4.8128 |
| 1850 | 176.75 | 0.00 | 0.00 | -3426.6 | 0.0 | 0.0 | 703 | 84 | -3497.7 | 5.0271 |
| 1900 | 187.75 | 0.00 | 0.00 | -3733.4 | 0.0 | 0.0 | 691 | 82 | -3806.4 | 5.2453 |
| 1950 | 198.75 | 0.00 | 0.00 | -4058.9 | 0.0 | 0.0 | 680 | 79 | -4133.9 | 5.4677 |
| 2000 | 210.25 | 0.00 | 0.00 | -4403.9 | 0.0 | 0.0 | 668 | 76 | -4480.8 | 5.6943 |

December 05, 2009

Exbal Ballistic Calculator

Federal Gold Medal: 308 Win. : Sierra MatchKing BTHP

| | SIGHT-IN DATA | FIELD DATA |
|---|---|---|
| Muzzle Velocity (fps) | 2600 | 2600 |
| Bullet Weight (grains) | 168 | |
| Sight Height (in) | 1.5 | |
| Sight-in Distance (yd) | 100 | |
| Altitude (ft) | 0 | 0 |
| Temperature (deg F) | 59 | 59 |
| Pressure@Sea Level(in Hg) | 29.53 | 29.53 |
| Relative Humidity (pct) | 78.0 | 78.0 |
| Wind Velocity (mph) | | 0 |
| Wind Angle (degrees) | | 180 ( 6.0 O'Clock) |
| Incline Angle (degrees) | | 0 |
| Moving Target Speed (mph) | | 0.0 |
| Ballistic Coefficient | 0.458 | |

POINT BLANK RANGE DATA

| TARGET HEIGHT (in) | SIGHT-IN DISTANCE (yd) | POINT BLANK RANGE HIGH (yd) | POINT BLANK RANGE LOW (yd) |
|---|---|---|---|
| 2 | 152 | 92 | 171 |
| 4 | 191 | 110 | 220 |
| 6 | 221 | 123 | 258 |
| 8 | 247 | 135 | 289 |
| 10 | 269 | 146 | 315 |

MAX APPARENT TRAJECTORY (in) 0.2

| TARGET DIST (yd) | SIGHT ADJUSTMENTS ELEV MOA | WIND MOA | LEAD MOA | TRAJECTORY VALUES ELEV (in) | WIND (in) | LEAD (in) | VELOCITY (fps) | ENERGY (ft-lb) | Drop from bore line (in) | ARRIVAL TIME (sec) |
|---|---|---|---|---|---|---|---|---|---|---|
| 0 | 0.00 | 0.00 | 0.00 | -1.5 | 0.0 | 0.0 | 2600 | 2521 | 0.0 | 0.0000 |
| 50 | 0.00 | 0.00 | 0.00 | -0.1 | 0.0 | 0.0 | 2504 | 2339 | -0.6 | 0.0588 |
| 100 | 0.00 | 0.00 | 0.00 | -0.0 | 0.0 | 0.0 | 2410 | 2166 | -2.7 | 0.1198 |
| 150 | 1.00 | 0.00 | 0.00 | -1.4 | 0.0 | 0.0 | 2318 | 2004 | -6.2 | 0.1832 |
| 200 | 2.25 | 0.00 | 0.00 | -4.5 | 0.0 | 0.0 | 2228 | 1852 | -11.3 | 0.2492 |
| 250 | 3.50 | 0.00 | 0.00 | -9.3 | 0.0 | 0.0 | 2140 | 1708 | -18.2 | 0.3179 |
| 300 | 5.00 | 0.00 | 0.00 | -16.0 | 0.0 | 0.0 | 2055 | 1574 | -27.0 | 0.3894 |
| 350 | 6.75 | 0.00 | 0.00 | -24.8 | 0.0 | 0.0 | 1971 | 1449 | -37.9 | 0.4639 |
| 400 | 8.50 | 0.00 | 0.00 | -35.7 | 0.0 | 0.0 | 1890 | 1332 | -51.0 | 0.5416 |
| 450 | 10.50 | 0.00 | 0.00 | -49.2 | 0.0 | 0.0 | 1810 | 1222 | -66.5 | 0.6227 |
| 500 | 12.50 | 0.00 | 0.00 | -65.3 | 0.0 | 0.0 | 1733 | 1120 | -84.6 | 0.7073 |
| 550 | 14.50 | 0.00 | 0.00 | -84.2 | 0.0 | 0.0 | 1658 | 1025 | -105.7 | 0.7958 |
| 600 | 17.00 | 0.00 | 0.00 | -106.3 | 0.0 | 0.0 | 1586 | 938 | -129.9 | 0.8882 |
| 650 | 19.50 | 0.00 | 0.00 | -131.9 | 0.0 | 0.0 | 1518 | 859 | -157.6 | 0.9849 |
| 700 | 22.00 | 0.00 | 0.00 | -161.3 | 0.0 | 0.0 | 1452 | 787 | -189.0 | 1.0859 |
| 750 | 24.75 | 0.00 | 0.00 | -194.8 | 0.0 | 0.0 | 1390 | 720 | -224.6 | 1.1915 |
| 800 | 27.75 | 0.00 | 0.00 | -232.7 | 0.0 | 0.0 | 1331 | 661 | -264.6 | 1.3017 |
| 850 | 31.00 | 0.00 | 0.00 | -275.6 | 0.0 | 0.0 | 1278 | 609 | -309.6 | 1.4168 |
| 900 | 34.25 | 0.00 | 0.00 | -323.8 | 0.0 | 0.0 | 1228 | 562 | -359.9 | 1.5366 |
| 950 | 38.00 | 0.00 | 0.00 | -377.7 | 0.0 | 0.0 | 1182 | 521 | -415.9 | 1.6611 |
| 1000 | 41.75 | 0.00 | 0.00 | -437.9 | 0.0 | 0.0 | 1142 | 486 | -478.2 | 1.7903 |
| 1050 | 46.00 | 0.00 | 0.00 | -504.8 | 0.0 | 0.0 | 1106 | 456 | -547.1 | 1.9239 |
| 1100 | 50.25 | 0.00 | 0.00 | -578.7 | 0.0 | 0.0 | 1075 | 431 | -623.1 | 2.0615 |
| 1150 | 54.75 | 0.00 | 0.00 | -660.2 | 0.0 | 0.0 | 1046 | 408 | -706.7 | 2.2032 |
| 1200 | 59.75 | 0.00 | 0.00 | -749.6 | 0.0 | 0.0 | 1020 | 388 | -798.2 | 2.3485 |
| 1250 | 64.75 | 0.00 | 0.00 | -847.4 | 0.0 | 0.0 | 997 | 371 | -898.1 | 2.4974 |
| 1300 | 70.00 | 0.00 | 0.00 | -954.0 | 0.0 | 0.0 | 975 | 355 | -1006.7 | 2.6498 |
| 1350 | 75.75 | 0.00 | 0.00 | -1069.7 | 0.0 | 0.0 | 956 | 341 | -1124.6 | 2.8055 |
| 1400 | 81.50 | 0.00 | 0.00 | -1194.9 | 0.0 | 0.0 | 937 | 328 | -1251.9 | 2.9643 |
| 1450 | 87.50 | 0.00 | 0.00 | -1330.1 | 0.0 | 0.0 | 920 | 316 | -1389.2 | 3.1262 |
| 1500 | 94.00 | 0.00 | 0.00 | -1475.6 | 0.0 | 0.0 | 905 | 305 | -1536.8 | 3.2911 |
| 1550 | 100.50 | 0.00 | 0.00 | -1631.8 | 0.0 | 0.0 | 890 | 295 | -1695.1 | 3.4589 |
| 1600 | 107.25 | 0.00 | 0.00 | -1799.1 | 0.0 | 0.0 | 876 | 286 | -1864.4 | 3.6296 |
| 1650 | 114.50 | 0.00 | 0.00 | -1977.8 | 0.0 | 0.0 | 862 | 277 | -2045.2 | 3.8031 |
| 1700 | 121.75 | 0.00 | 0.00 | -2168.3 | 0.0 | 0.0 | 849 | 269 | -2237.8 | 3.9794 |
| 1750 | 129.50 | 0.00 | 0.00 | -2371.0 | 0.0 | 0.0 | 836 | 260 | -2442.6 | 4.1587 |
| 1800 | 137.25 | 0.00 | 0.00 | -2586.3 | 0.0 | 0.0 | 823 | 253 | -2659.9 | 4.3408 |
| 1850 | 145.25 | 0.00 | 0.00 | -2814.6 | 0.0 | 0.0 | 811 | 245 | -2890.4 | 4.5258 |
| 1900 | 153.50 | 0.00 | 0.00 | -3056.3 | 0.0 | 0.0 | 799 | 238 | -3134.2 | 4.7138 |
| 1950 | 162.25 | 0.00 | 0.00 | -3311.9 | 0.0 | 0.0 | 788 | 232 | -3391.9 | 4.9046 |
| 2000 | 171.00 | 0.00 | 0.00 | -3581.8 | 0.0 | 0.0 | 777 | 225 | -3663.8 | 5.0984 |

December 05, 2009

**.308 Winchester 168-grain Sierra Match King Bullet Ballistics**

Exbal Ballistic Calculator

Lapua Target: 308 Win : Scenar

| | SIGHT-IN DATA | FIELD DATA |
|---|---|---|
| Muzzle Velocity (fps) | 2690 | 2690 |
| Bullet Weight (grains) | 167 | |
| Sight Height (in) | 1.5 | |
| Sight-in Distance (yd) | 100 | |
| Altitude (ft) | 0 | 0 |
| Temperature (deg F) | 59 | 59 |
| Pressure@Sea Level(in Hg) | 29.53 | 29.53 |
| Relative Humidity (pct) | 78.0 | 78.0 |
| Wind Velocity (mph) | | 0 |
| Wind Angle (degrees) | | 180 ( 6.0 O'Clock) |
| Incline Angle (degrees) | | 0 |
| Moving Target Speed (mph) | | 0.0 |
| Ballistic Coefficient | 0.470 | |

POINT BLANK RANGE DATA

| TARGET HEIGHT (in) | SIGHT-IN DISTANCE (yd) | POINT BLANK RANGE HIGH (yd) | POINT BLANK RANGE LOW (yd) |
|---|---|---|---|
| 2 | 158 | 96 | 178 |
| 4 | 198 | 114 | 228 |
| 6 | 229 | 127 | 266 |
| 8 | 255 | 140 | 300 |
| 10 | 279 | 153 | 326 |

MAX APPARENT TRAJECTORY (in) 0.1

| TARGET DIST (yd) | SIGHT ADJUSTMENTS ELEV MOA | WIND MOA | LEAD MOA | TRAJECTORY VALUES ELEV (in) | WIND (in) | LEAD (in) | VELOCITY (fps) | ENERGY (ft-lb) | Drop from bore line (in) | ARRIVAL TIME (sec) |
|---|---|---|---|---|---|---|---|---|---|---|
| 0 | 0.00 | 0.00 | 0.00 | -1.5 | 0.0 | 0.0 | 2690 | 2683 | 0.0 | 0.0000 |
| 50 | 0.25 | 0.00 | 0.00 | -0.1 | 0.0 | 0.0 | 2595 | 2496 | -0.6 | 0.0568 |
| 100 | 0.00 | 0.00 | 0.00 | -0.0 | 0.0 | 0.0 | 2501 | 2320 | -2.5 | 0.1156 |
| 150 | 0.75 | 0.00 | 0.00 | -1.3 | 0.0 | 0.0 | 2410 | 2153 | -5.8 | 0.1767 |
| 200 | 2.00 | 0.00 | 0.00 | -4.1 | 0.0 | 0.0 | 2320 | 1995 | -10.6 | 0.2401 |
| 250 | 3.25 | 0.00 | 0.00 | -8.5 | 0.0 | 0.0 | 2232 | 1847 | -16.9 | 0.3060 |
| 300 | 4.75 | 0.00 | 0.00 | -14.6 | 0.0 | 0.0 | 2146 | 1708 | -25.1 | 0.3745 |
| 350 | 6.25 | 0.00 | 0.00 | -22.6 | 0.0 | 0.0 | 2063 | 1578 | -35.1 | 0.4458 |
| 400 | 7.75 | 0.00 | 0.00 | -32.7 | 0.0 | 0.0 | 1981 | 1455 | -47.2 | 0.5199 |
| 450 | 9.50 | 0.00 | 0.00 | -45.0 | 0.0 | 0.0 | 1902 | 1341 | -61.4 | 0.5972 |
| 500 | 11.50 | 0.00 | 0.00 | -59.6 | 0.0 | 0.0 | 1824 | 1233 | -78.1 | 0.6777 |
| 550 | 13.25 | 0.00 | 0.00 | -76.9 | 0.0 | 0.0 | 1748 | 1133 | -97.4 | 0.7617 |
| 600 | 15.50 | 0.00 | 0.00 | -97.0 | 0.0 | 0.0 | 1675 | 1040 | -119.5 | 0.8493 |
| 650 | 17.75 | 0.00 | 0.00 | -120.3 | 0.0 | 0.0 | 1604 | 954 | -144.7 | 0.9408 |
| 700 | 20.00 | 0.00 | 0.00 | -146.9 | 0.0 | 0.0 | 1536 | 875 | -173.4 | 1.0364 |
| 750 | 22.50 | 0.00 | 0.00 | -177.2 | 0.0 | 0.0 | 1472 | 803 | -205.6 | 1.1361 |
| 800 | 25.25 | 0.00 | 0.00 | -211.5 | 0.0 | 0.0 | 1410 | 737 | -241.9 | 1.2402 |
| 850 | 28.00 | 0.00 | 0.00 | -250.1 | 0.0 | 0.0 | 1351 | 677 | -282.6 | 1.3489 |
| 900 | 31.25 | 0.00 | 0.00 | -293.6 | 0.0 | 0.0 | 1297 | 624 | -328.0 | 1.4622 |
| 950 | 34.50 | 0.00 | 0.00 | -342.1 | 0.0 | 0.0 | 1247 | 577 | -378.6 | 1.5801 |
| 1000 | 37.75 | 0.00 | 0.00 | -396.3 | 0.0 | 0.0 | 1201 | 535 | -434.8 | 1.7026 |
| 1050 | 41.50 | 0.00 | 0.00 | -456.5 | 0.0 | 0.0 | 1159 | 498 | -497.0 | 1.8298 |
| 1100 | 45.50 | 0.00 | 0.00 | -523.2 | 0.0 | 0.0 | 1123 | 467 | -565.6 | 1.9614 |
| 1150 | 49.50 | 0.00 | 0.00 | -596.7 | 0.0 | 0.0 | 1090 | 441 | -641.2 | 2.0971 |
| 1200 | 54.00 | 0.00 | 0.00 | -677.6 | 0.0 | 0.0 | 1061 | 417 | -724.0 | 2.2367 |
| 1250 | 58.50 | 0.00 | 0.00 | -766.2 | 0.0 | 0.0 | 1034 | 397 | -814.6 | 2.3800 |
| 1300 | 63.50 | 0.00 | 0.00 | -862.9 | 0.0 | 0.0 | 1010 | 379 | -913.4 | 2.5269 |
| 1350 | 68.50 | 0.00 | 0.00 | -968.2 | 0.0 | 0.0 | 988 | 362 | -1020.6 | 2.6773 |
| 1400 | 73.75 | 0.00 | 0.00 | -1082.4 | 0.0 | 0.0 | 968 | 348 | -1136.8 | 2.8309 |
| 1450 | 79.50 | 0.00 | 0.00 | -1205.9 | 0.0 | 0.0 | 949 | 334 | -1262.3 | 2.9877 |
| 1500 | 85.25 | 0.00 | 0.00 | -1339.0 | 0.0 | 0.0 | 932 | 322 | -1397.5 | 3.1476 |
| 1550 | 91.25 | 0.00 | 0.00 | -1482.3 | 0.0 | 0.0 | 916 | 311 | -1542.7 | 3.3105 |
| 1600 | 97.75 | 0.00 | 0.00 | -1635.9 | 0.0 | 0.0 | 901 | 301 | -1698.3 | 3.4762 |
| 1650 | 104.25 | 0.00 | 0.00 | -1800.4 | 0.0 | 0.0 | 887 | 291 | -1864.8 | 3.6447 |
| 1700 | 111.00 | 0.00 | 0.00 | -1975.9 | 0.0 | 0.0 | 873 | 282 | -2042.4 | 3.8160 |
| 1750 | 118.00 | 0.00 | 0.00 | -2163.0 | 0.0 | 0.0 | 859 | 274 | -2231.5 | 3.9901 |
| 1800 | 125.25 | 0.00 | 0.00 | -2362.0 | 0.0 | 0.0 | 847 | 266 | -2432.4 | 4.1670 |
| 1850 | 132.75 | 0.00 | 0.00 | -2573.3 | 0.0 | 0.0 | 834 | 258 | -2645.7 | 4.3467 |
| 1900 | 140.50 | 0.00 | 0.00 | -2797.2 | 0.0 | 0.0 | 822 | 251 | -2871.6 | 4.5292 |
| 1950 | 148.50 | 0.00 | 0.00 | -3034.2 | 0.0 | 0.0 | 810 | 244 | -3110.6 | 4.7146 |
| 2000 | 156.75 | 0.00 | 0.00 | -3284.7 | 0.0 | 0.0 | 799 | 237 | -3363.1 | 4.9027 |

December 05, 2009

**.308 Winchester 167 grain Lapua Scenar bullet ballistics**

# Acknowledgments

The authors basically had no idea what they were getting into when coming up with the idea to write this book. Many people have helped along the way and deserve thanks, appreciation, and recognition for their time and effort. Dr. Curtis Prejean is a good friend and provided incredible content for us. He is a virtual encyclopedia of knowledge with regards to weapons and ballistics, and just an all around great guy to boot. Thanks to Matt Johnson for getting out of his banker's chair to find some great old photos for us to use in the book, and to Scott Tyler for his stories and photos that he was willing to donate to the cause. To Marc Halcon of American Shooting Center here in San Diego for his network, his knowledge, and access to all his great equipment. Thank you also to our good friend and fellow pirate Billy Tosheff and tech genius Thomas Frasher for his excellent input. We are definitely going to be buying beers for Tony Lyons and Jay McCullough, our publisher and editor respectively, for their incredible patience with a couple of complete rookies who like to put too much on their plates. Lastly but surely not the least, to Lana Kizy, for all her tireless help and for putting up with our BS.